Low Speed Marine Diesel

Ocean Engineering: A Wiley Series

EDITOR:
MICHAEL E. McCORMICK, Ph.D.
U.S. Naval Academy

ASSOCIATE EDITOR:
RAMESWAR BHATTACHARYYA, D. Ingr.
U.S. Naval Academy

Michael E. McCormick	Ocean Engineering Wave Mechanics
John B. Woodward	Marine Gas Turbines
H. O. Berteaux	Buoy Engineering
Clarence S. Clay and Herman Medwin	Acoustical Oceanography: Principles and Applications
F. W. Wheaton	Aquacultural Engineering
Robert M. Sorensen	Basic Coastal Engineering
Rameswar Bhattacharyya	Dynamics of Marine Vehicles
Stephen C. Dexter	Handbook of Oceanographic Engineering Materials
John B. Woodward	Low Speed Marine Diesel

LOW SPEED MARINE DIESEL

JOHN B. WOODWARD

Department of Naval Architecture and Marine Engineering
University of Michigan
Ann Arbor, Michigan

A WILEY-INTERSCIENCE PUBLICATION
JOHN WILEY & SONS, New York · Chichester · Brisbane · Toronto

Copyright © 1981 by John Wiley & Sons, Inc.

All rights reserved. Published simultaneously in Canada.

Reproduction or translation of any part of this work
beyond that permitted by Sections 107 or 108 of the
1976 United States Copyright Act without the permission
of the copyright owner is unlawful. Requests for
permission or further information should be addressed to
the Permissions Department, John Wiley & Sons, Inc.

Library of Congress Cataloging in Publication Data:

Woodward, John B.
  Low speed marine diesel.

  (Ocean engineering)
  "A Wiley-Interscience publication."
  Includes bibliographies and index.
  1. Marine diesel motors.   I. Title.   II. Series:
Ocean engineering series.

VM770.W66.   623.8'7236   80-39635
ISBN 0-471-06335-5

Printed in the United States of America

10 9 8 7 6 5 4 3 2 1

SERIES PREFACE

Ocean engineering is both old and new. It is old in that man has concerned himself with specific problems in the ocean for thousands of years. Ship building, prevention of beach erosion, and construction of offshore structures are just a few of the specialties that have been developed by engineers over the ages. Until recently, however, these efforts tended to be restricted to specific areas. Within the past decade an attempt has been made to coordinate the activities of all technologists in ocean work, calling the entire field "ocean engineering." Here we have its newness.

Ocean Engineering: A Wiley Series has been created to introduce engineers and scientists to the various areas of ocean engineering. Books in this series are so written as to enable engineers and scientists easily to learn the fundamental principles and techniques of a specialty other than their own. The books can also serve as textbooks in advanced undergraduate and introductory graduate courses. The topics to be covered in this series include ocean engineering wave mechanics, marine corrosion, coastal engineering, dynamics of marine vehicles, offshore structures, and geotechnical or seafloor engineering. We think that this series fills a great need in the literature of ocean technology.

                                                      MICHAEL E. MCCORMICK, EDITOR
                                        RAMESWAR BHATTACHARYYA, ASSOCIATE EDITOR

November 1972

PREFACE

The diesel engine comes close to being the universal Prime Mover, suited to all demands for power generation. Within the limited field of commercial marine propulsion it could indeed be the one type of engine used for all vessels. True, the gas turbine gained a brief foothold in the early 1970s, and the steam turbine may continue to hold a small share of the propulsion market, but the advantages these rivals may show over diesel in their few applications are quite small. All commercial vessels *could* be diesel with no disadvantage in the great majority of cases, and only modest disadvantage in the remainder.

However, there is a distinct gap in this expansive domain: a prominent member of the diesel family—the low speed engine that is direct connected (connected to the propeller without intervening reduction gears)—has been totally excluded from the U. S. merchant marine by a firm traditional adherence to steam propulsion. When I began this book in 1977, no such engine had been installed by an American shipyard since before World War II. On the other hand, the European builders of low speed marine engines have been vigorously promoting the undoubted virtues of their product for many years, and have provided the power for most of the world's large merchant ships. It was clear in 1977 that they were on the verge of some success in America also and, as this book progressed in 1978 and 1979, that they had broken into this last stronghold of resistance to low speed diesel propulsion. For example, two product carriers (petroleum tankers) powered by Sulzer low speed engines were launched in 1979 at the Sun Shipbuilding plant in Chester, Pennsylvania, apparently a prophetic first. News also came of others, and even of conversion of existing steamers to diesel (though not all involved low speed engines).

PREFACE

The principal cause of the sudden surge of the low speed diesel into new territory is the steep rise of fuel prices in the 1970s, for the ability of this engine to make efficient use of fuel drawn from the low end of the price spectrum is unsurpassed. Since the trends in fuel price appear irreversible, it is likely that the low speed diesel engine will be an important factor, perhaps a dominating one, in the American merchant marine, just as it has been worldwide.

Familiarity with the low speed diesel now becomes essential for the American shipbuilders, ship designers, shipowners, and the students who are preparing to join them, since they can no longer equate "ship engine" with "steam turbine." This book is written with them in mind, particularly the engineering student, who can use it not only while in school, but later as a reference.

Description of the engines is certainly necessary in a book like this: however, the major thrust is not description, but the understanding needed for application of the engines to ship propulsion. Descriptions can be gotten from literature that is freely distributed by the engine builders, and that source is preferred, for details vary with builders and with design changes. Likewise, treatment of diesel principles, even thermodynamics, is essential, and some of this is to be found here, but once again, the major theme is the ship design application. In short, the book is aimed at those who must understand the propulsion aspects of ship design—the student designer in particular—and is intended to fill the gap between a text on diesel theory and principles and the purely descriptive engine literature.

The engine builders are the major source of data quoted here. Sulzer, MAN, Burmeister & Wain, Grandi Motori Trieste, Doxford, Mitsubishi, and several licensees of these companies build low speed engines, and generously supply information when requested. You will note particularly that Sulzer engines are the ones mentioned the most frequently in text and illustrations. This is due not to favoritism, but simply to the interest generated during the writing period by the building of ships with Sulzer engines at Sun Shipbuilding. Thanks go to all of the manufacturers for their contributions, but perhaps a little extra gratitude is deserved by Sulzer and by Ernst Jung, their manager in New York. Sun Shipbuilding and their manager of engineering, Hector McVey, were also especially helpful.

JOHN B. WOODWARD

Ann Arbor, Michigan
January 1981

CONTENTS

1. Introduction   1

1.1 Introduction to the Low Speed Engine, 4
1.2 The Marine Designer's Introduction, 6
1.3 Notation for Chapter 1, 9

2. General Engine Principles   10

2.1 The Diesel Cycle, 12
2.2 The PLAN Formula and Engine Torque and Power Characteristics, 14
2.3 Piston Speed, 18
2.4 Basic Engine Dimensions, 19
2.5 More on Mean Effective Pressure, 20
2.6 Stroke Length, 23
2.7 Summary of Engine Limitations, 24
2.8 Notation for Chapter 2, 25

3. Engine Construction   27

3.1 Pistons and Running Gear, 27
3.2 Crankshaft, 35
3.3 Cylinder Block and Cylinder Liner, 37
3.4 Cylinder Covers, 39
3.5 General Structure, 39
3.6 Cylinder Ports and Valves, 40
3.7 Fuel Pumps and Injectors, 45
3.8 Air and Exhaust, 49

3.9   Lubrication, 51
3.10 Cooling, 51
3.11 Classification Society Rules on Engine Construction, 52
3.12 References, 53
3.13 Notation for Chapter 3, 53

4. Engine Operational Characteristics — 54

4.1 Torque and Power Characteristics, 55
4.2 Fuel Consumption Characteristics, 55
4.3 General Engine Characteristics, 58
4.4 Engine Heat Balance, 60
4.5 Effect of Ambient Conditions on Some Characteristics, 60
4.6 Turbocharger and Engine Air Flow Characteristics, 64
4.7 References, 64
4.8 Notation for Chapter 4, 64

5. The Relationship of Engine to Propeller — 65

5.1 Propeller Characteristics, 67
5.2 Torque and Power Characteristics of Propellers, 69
5.3 Basic Match of Engine and Propeller Characteristics, 71
5.4 Design for Resistance Change, 74
5.5 Controllable Pitch Propellers, 75
5.6 Nonpropulsive Loads, 78
5.7 Transient Conditions, 80
5.8 References, 82
5.9 Notation for Chapter 5, 82

6. Scavenging and Turbocharging — 84

6.1 Scavenging Principles, 85
6.2 Scavenging Arrangements, 87
6.3 General Principles of Turbocharging, 89
6.4 Turbocharger Characteristics, 94
6.5 Pulse Operation and Constant Pressure Operation, 97
6.6 Turbochargers Combined with Mechanically Driven Blowers, 100
6.7 Two-Stage Turbocharging, 104
6.8 References, 106
6.9 Notation for Chapter 6, 107

7. Engine Rating — 108

7.1 Factors in the Rating, 109

7.2 Rating Corrections, 112
7.3 Choosing RPM, 114
7.4 Operating Margin, 114
7.5 Hull Service Margin, 114
7.6 Service Power and Rating, 116
7.7 Comparison Criteria, 118
7.8 References, 121
7.9 Notation for Chapter 7, 122

8. Fuels and Lubricants 123

8.1 Fuel Properties, 125
8.2 Fuel Specifications, 129
8.3 Fuels for the Low Speed Diesel, 130
8.4 Fuel Treatment, 136
8.5 Lubricants, 137
8.6 Gaseous Fuel, 138
8.7 References, 140
8.8 Appendix—Some Things about Viscosity, 142
8.9 Notation for Chapter 8, 143

9. Control and Monitoring 144

9.1 Engine Control, 145
9.2 Computer Control, 147
9.3 Engine Control from Remote Locations, 149
9.4 Engine Monitoring, 162
9.5 Control with a Controllable Pitch Propeller, 167
9.6 Fuel Control with Liquefied Natural Gas Fuel, 169
9.7 References, 170
9.8 Appendix—Governing, 171
9.9 Notation for Chapter 9, 178

10. Engine Auxiliary Systems 179

10.1 Introduction, 179
10.2 Engine Cooling Systems, 180
10.3 Fuel System, 188
10.4 Lubricating Oil Systems, 191
10.5 Starting Air, 194
10.6 Ventilation Air, 196
10.7 Exhaust, 198
10.8 Exhaust Gas Waste Heat, 200
10.9 Electric Load for Engine Auxiliaries, 204

xii CONTENTS

10.10 Machinery Weight, 206
10.11 Shaft Drive for Auxiliary Power, 207
10.12 Fire Protection, 208
10.13 Cleaning Systems, 210
10.14 References, 210
10.15 Notation for Chapter 10, 210

11. Engine Failures, Wear, and Maintenance   **214**

11.1 Introduction, 214
11.2 Abrupt Degradations, 215
11.3 Expected Maintenance, 219
11.4 Maintenance Equipment, 225
11.5 Spare Parts, 230
11.6 Major Repairs in situ, 231
11.7 References, 232
11.8 Notation for Chapter 11, 233

12. The Engine and Its Environment   **234**

12.1 Introduction, 234
12.2 Engine Room Arrangement, 235
12.3 Engine Mounting in Ship, 236
12.4 Torsional Vibration of Engine and Shaft, 246
12.5 Axial Shaft Vibration, 255
12.6 Engine Inertia Forces and Torques, 256
12.7 Engine Lateral Excitations, 258
12.8 Summary of Dynamic Problems, 259
12.9 Noise, 259
12.10 The Engine and Its Environment beyond the Ship, 265
12.11 References, 266
12.12 Notation for Chapter 12, 267

Index   **269**

Low Speed Marine Diesel

Chapter One

INTRODUCTION

The diesel engine comes close to being the universal marine propulsion engine. Although it has superior competition for some types of propulsion, it can be used successfully in almost any application, and it is generally acknowledged to be the best in a wider range of marine applications than any other engine.

The attribute contributing the largest share to the accolade 'best' is efficiency. Diesel specific fuel consumption in most applications is lower (synonymous with higher efficiency) than either steam or gas turbine or spark ignition (gasoline) plants, and the fuel is usually at least as cheap per unit of heating value as theirs. Compared to the steam propulsion plant, the diesel also enjoys the advantages of internal combustion, which makes it compact, available in essentially a complete package, and simple to control. The gas turbine and spark-ignition engines are also internal combustion, of course, but the efficiency advantage heavily favors diesel over both of these. The use of a fuel of much lower volatility than gasoline also gives it an important safety advantage over spark ignition.

But perhaps the easiest way to picture the broad applicability of the diesel engine is to delineate those few areas where it is *not* clearly superior to the alternatives.

Its most obvious area of competitive disadvantage is in the small pleasure boat field, dominated by the spark-ignition engine. The latter engine is distinctly lower in first cost, and is also of lighter weight. Its greater

2 INTRODUCTION

consumption of a more expensive fuel is of small consequence in the typical lightly used pleasure boat. Hence millions of small boats are powered by the spark-ignition engine. Nonetheless, diesels are frequently used for pleasure boats that are designed for long cruising range, since low specific fuel consumption means greater range for a given tankage. Diesel fuel is also significantly cheaper than gasoline in some neighborhoods, and is definitely less hazardous in enclosed spaces.

Another area of disadvantage is at the opposite end of the power range, that occupied by large and fast ships, or by ships so large that they require many thousands of horsepower even at modest speeds. During the 1960s and 1970s, propulsion powers above about 30,000 shp were dominated by the steam plant. (A very rough boundary, it varied with many factors that shall not be discussed here.) For these high powers, the high steam conditions and the complexity of regenerative arrangements that an efficient steam plant needs seem economically attractive. The efficiency is almost equal to that of diesel, and the lower lubricating oil consumption, lower weight, and possibly lower maintenance costs make the steam alternative look much better than at lower powers. Both competitors burn essentially the same fuel. The diesel suffers at high powers because its cylinder displacement, and hence engine size and weight, increase faster than power rating. An alternative is to seek high power through use of many cylinders, but this route leads to excessive complexity.*

The gas turbine is another successful competitor in applications where compactness and lightness are essential. Thus it is used in some high speed vessels whose payload would be drastically cut by machinery weight if a propulsion plant of the required high power were to be diesel.

The exceptions mentioned include only a few of the world's merchant ships, and exceptions aside, the commercial propulsion field is dominated by the diesel. Table 1.1 is an indication of just how strong this dominance is. Although both gas turbine and steam technologies are subject to continued improvement, projections forward from 1980 indicate that neither can reverse this general dominance of diesel, until (and unless) there is a radical shift in the type or quality of fuel available for marine engines.

This book is concerned especially with the engines that propel the large oceangoing merchant ship, a part of the marine universe that is dominated not just by the diesel, but often by a particular member of the diesel family: the low speed, direct connected engine.

---

*The boundary between choice of steam and choice of diesel depends strongly on the relative importances of the factors mentioned. In 1980, for example, high fuel prices greatly increased the importance of efficiency, and so magnified the effect of the small efficiency margin of the diesel over the large steam plants. At that time, therefore, diesel propulsion was encroaching on territory formerly dominated by steam.

TABLE 1.1 PROPULSION MACHINERY, SHIPS COMPLETED IN 1975

Type of Engine	Ships	Number of Engines	Total Horsepower	Percent of Horsepower
Low speed diesel	477	482	7220560	49.2
Medium speed diesel	411	621	2848370	19.4
Steam	130	132	4585000	31.2
Gas turbine	2	2	24600	0.16

(*Source: THE MOTOR SHIP* Journal, January 1976.)

It is a particular member, indeed, for all diesel engines are not alike. Far from it. To cover the extensive and varied propulsion field, an extensive and varied array of engine types is necessary. There are a number of ways of classifying them. For a first example, marine diesel engines are either in-line or V—a classification of their cylinder arrangements—or they are either two-stroke or four-stroke—a classification by the internal combustion cycle used. More important for the purposes of this discussion is classification into *high speed*, *medium speed*, and *low speed* categories, a classification most conveniently based on the rated speed (rpm).

High speed engines have rated speeds in the range 1000 to 2400 rpm,* most typically are adaptations of truck or construction-machinery engines, and power smaller vessels (less than 1000 hp per engine). Medium speed engines have rated speeds in the range 400 to 900 rpm, with power ratings in the 1000 to 10,000 hp range. Medium speed engines of North American manufacture are usually adaptations of engines built for electric power generation ashore; the engines at the higher end of the medium speed rpm range are diesel-electric locomotive engines. The medium speed engines are the usual power for the 'middle-sized' vessels—tugs, river towboats, ferries, larger fishing vessels, etc. However, larger vessels can also be powered by combining several of these engines. The Great Lakes 'thousand-footer' bulk carriers built after 1970 are a prominent American example; many of them obtain 14,000 to 19,000 shp from combination of two or four medium speed engines. Low speed engines have rated speeds in the range 100 to 150 rpm, with power ratings covering the range 5000 to 40,000 hp. These are the true ship propulsion engines, the engines most commonly used to propel the oceangoing

---

*The divisions stated here between classes of engines are somewhat arbitrary; other authors might locate them differently. For example, locomotive engines rated at 900 rpm are called 'high speed' by some ship people.

4 INTRODUCTION

merchant ship. Note that their speed range is also that of a ship propeller, and indeed they are designed with this in mind. Therefore they, unlike high speed and medium speed engines, do not require a speed-reducing gear between their output coupling and the propeller shaft.

1.1 INTRODUCTION TO THE LOW SPEED ENGINE

This book is about ship propulsion by the low speed diesel engine. "Ship" implies a large oceangoing vessel (possibly Great Lakes also), typical of the many and diversifed units of the commercial fleets of most maritime nations.

To introduce the low speed ship propulsion engine, one should observe and explain the terminology that especially applies to the engine of interest. *Low speed* has already been noted. This engine is further characterized by the terms *direct connected*, *two-stroke*, *turbocharged*, *crosshead*, *in-line*, and *cathedral* (colloquial, and even faintly pejorative, this last).

Direct connected calls attention to one of the attractive consequences of low speed, that of direct connection (no intervening gear set or other speed reducer) between engine coupling and propulsion shaft.

Two-stroke denotes the completion of a cylinder cycle—intake, compression, ignition, and exhaust—over two strokes of the piston. The alternative, quite common in smaller engines, is the four-stroke cycle in which each of the four cylinder events requires a stroke of the piston. Because the two-stroke engine obtains twice as many power strokes per cycle as the four-stroke alternative, it nominally can produce twice the power from a given displacement (engine size) and rotational speed. Other factors more subtle greatly reduce the magnitude of this advantage, but enough remains that a two-stroke engine can be somewhat lower in specific weight (weight per unit of power) than four-stroke engines. The low speed engine is inevitably a large engine (for reasons to be explained in the next chapter), and so must take advantage of any factor that tends toward size reduction, this to avoid being just too massive for practicality. All low speed marine engines are therefore two-stroke engines.

Turbocharged implies the use of superchargers driven by exhaust gas turbines (turbine + supercharger = turbocharger) to supply combustion air at pressure above atmospheric. This feature is used by diesel engines of all sizes and types to increase power output from a given cylinder volume. The exhaust-driven turbocharger can be seen in the cross-sectional view of a low speed engine (Figure 1.1).

Crosshead refers to the universal use by this engine type of crosshead construction for the reciprocating assembly (piston and its connection to

INTRODUCTION TO THE LOW SPEED ENGINE

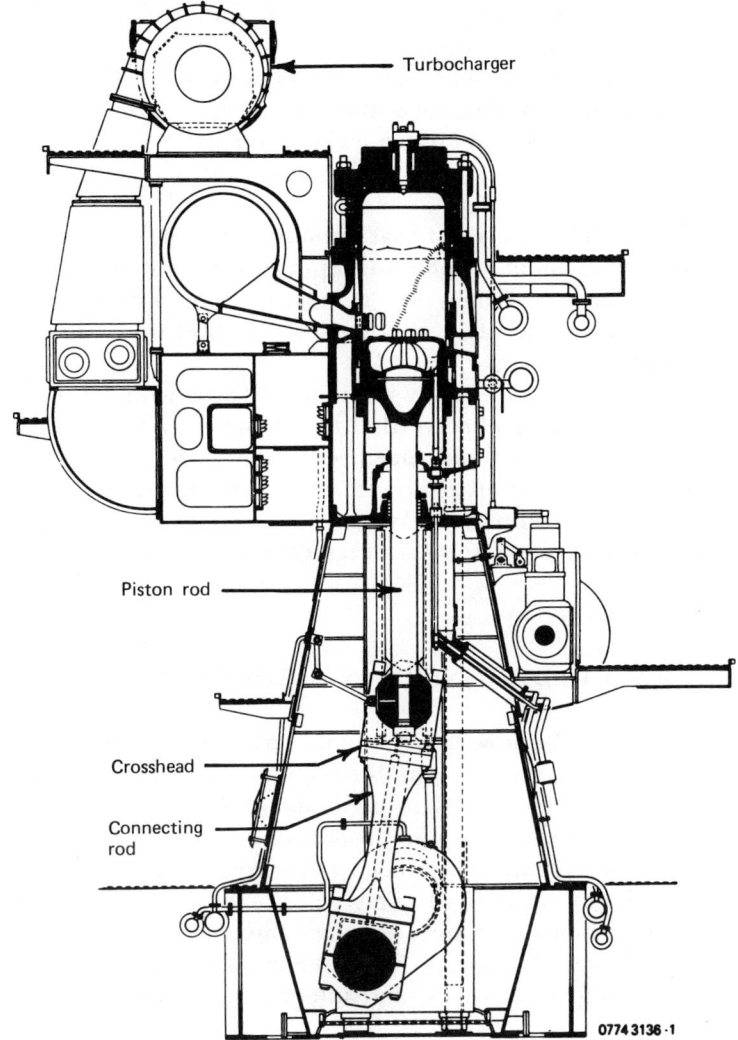

Figure 1.1 Cross section of low speed ship propulsion engine, Sulzer RND 105.

the crankshaft). The crosshead is labeled in Figure 1.1. It is connected to the piston by a piston rod whose motion is precisely linear. A seal around this rod is thus feasible, thereby permitting sealing of the crankcase from the space beneath the pistons, a feature that is necessary to prevent contamination of crankcase oil by cooling water leaking from the pistons. (Noncrosshead, or *trunk piston* engines, use oil as the piston coolant, or have uncooled pistons.) The sealed space under the piston can also serve as an air compression chamber, and is often used as such to supplement the turbocharger.

In line distinguishes the cylinder arrangement from the V arrangement that is common in medium speed and high speed engines. The large piston mass of the low speed engine makes a vertical cylinder configuration essential, in turn making a vertical in-line array of cylinders the only practicable one.

Cathedral is an informal term that connotes the large size (either in awe or in disparagement, depending upon the user's feelings toward this type of engine). In either case it must certainly be the consensus that a low speed engine of high power, say 10,000 hp or more, is of impressive size.

To summarize the introductory points made here with respect to the low speed engine, let us observe that it is a two-stroke, turbocharged, in-line engine, coupled without speed reduction to the propeller. It employs crosshead construction in its reciprocating elements, and is the only type to use water-cooled pistons. Worldwide, it is the favorite for ocean-going commercial ship propulsion. A typical example is pictured in Figure 1.2.

1.2 THE MARINE DESIGNER'S INTRODUCTION

An introduction to the low speed engine has just been given; further description of this engine appears throughout the remaining chapters. Meanwhile, let it be observed that the theme of this book is the application of the low speed diesel to ship propulsion. The intended readership is the design oriented marine engineer (hereafter 'marine designer, 'ship designer,' or simply 'designer'). This will not be the person who will design the *engine*, nor operate it, nor maintain it. Rather, it will be the person whose basic task lies external to the engine, a task of choosing it wisely from among alternatives, and integrating it into a harmonious propulsion plant.

The choice of engine is somewhat beyond the scope of this book, since the general implication is use of low speed diesel. On the other hand, this engine is not the best choice for all applications, making it unwise for a marine designer to work unaware of the features that make it suitable or unsuitable for an application at hand. Here, briefly, let us note that these questions concern features that determine the weight and volume requirements of the machinery, the many categories of cost, and the effect of several intangible interactions with the ship (for example, excitation of vibration).

THE MARINE DESIGNER'S INTRODUCTION 7

Figure 1.2 Low speed ship propulsion engine, Sulzer RND 105.

Integration into the propulsion plant requires major effort in ship design, a greater effort than might be suspected from the "package" nature of the engine. The first consideration is interaction with the load, namely, the propeller. The propeller must be designed to meet requirements of the hull and hence cannot be part of the engine package, yet propeller and engine form a load and driver pair that must match the respective characteristics of each. Other loads are often present also. For instance, engine exhaust gas is often used as the heat source in an auxiliary boiler ('waste heat boiler'), and sometimes as a source of inert gas

8 INTRODUCTION

for suppressing the explosive vapors in cargo oil tanks. Although these loads do not interact with the engine in the direct manner of a mechanical load, the designer must at least know of the engine's characteristics in order to determine those of the dependent systems.

Secondary mechanical loads are sometimes found. The most likely example is a shaft-driven electrical generator. Such loads participate in the propeller-engine interaction, thus adding another dimension to propeller design.

The propeller must be driven by a shaft. Although the low speed engine propulsion plant does not use a reduction gear in its transmission line, the shaft and its appurtenances alone demand significant engineering effort. Bearings and seals always demand design attention, and the shaft-engine-propeller array constitutes a mass-elastic system that is subject to vibrations of several kinds. Because of the inevitable pulsations in the torque of a diesel engine, torsional vibration is the kind most likely to be encountered, and is a threat to be analyzed in every design. A torsional vibration analysis of the transmission array is consequently a necessity.

The engine must be supported by a foundation. Although one may say that foundations are structure, not machinery, their design is influenced not only by mere weight of the engine, but by other characteristics as well.

Then there are the many auxiliary services. Engines of all types require lubrication, fuel supply, control and monitoring, starting, air supply and exhaust removal. Part of all of these are typically included in the diesel engine 'package,' but the engine builder cannot provide the auxiliary systems in toto. For example, a low speed engine requires water cooling of its pistons,* its cylinder and cylinder head jackets, its lube oil, and its combustion air. The several heat exchangers and pumps are likely to be integrated with the engine by the builder, but even so, the ship designer will be faced with design of the piping that connects engine systems to the sea. In general, all engine auxiliary systems require design participation by the ship designer.

In summary, let it then be noted that the process from selection of engine through design of the overall propulsion plant is an engineering task of many parts. Although none may involve work on the engine itself, all require understanding of this principal component and its characteristics. The following chapters therefore have the purpose of helping the marine designer with this understanding.

---

*Water cooling of pistons is mentioned several times in this chapter, but it should be noted that this is not universal practice, since some engine builders use oil cooling.

1.3 NOTATION FOR CHAPTER 1

bhp   brake horsepower
hp   horsepower
rpm   rotational speed, per minute
shp   shaft horsepower

Chapter

Two

GENERAL ENGINE PRINCIPLES

The principal applied concepts that distinguish the diesel engine from other types can be summed in a few words: it is a reciprocating, internal combustion, compression ignition engine.

It is assumed here that even the novice does not require discussion of the first two terms, since the many millions of commonplace gasoline-fueled engines (for example, automotive engines) are also reciprocating internal combustion machines. Compression ignition, on the other hand, is unique to the diesel, and so demands space in any comprehensive discussion of this engine type.

Put simply, the ignition of fuel in a diesel cylinder is accomplished by compression of the air in the cylinder to a temperature above fuel ignition temperature. The process can be described in more detail with the aid of Figure 2.1, which is a cross-sectional view of a cylinder with the piston near its bottom center position. Air supplied from an external source (not shown) is entering through the intake ports. The upward stroke of the piston will first close these ports, then compress the trapped air as it proceeds toward top center. Air pressure reaches from 60 to 75 kPa, which puts air temperature in the neighborhood of 500 C as a result of the compression. At a point somewhat before the maxima are reached, fuel (a liquid petroleum fuel) is sprayed into the cylinder via the *injector*, and after a brief delay in which the fuel is heated and vaporized by its environment, it ignites and burns rapidly. The release of energy drives

GENERAL ENGINE PRINCIPLES 11

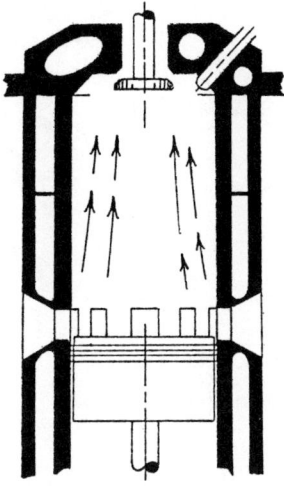

Figure 2.1 Cross section of diesel cylinder. Air shown entering through inlet ports.

pressure and temperature much higher, and as the piston proceeds into the down stroke, provides the mechanical energy input to the piston, and hence to the crankshaft and its load. Exhaust takes place while the piston is near bottom center. The exhaust valve is open briefly so that the spent gas can flow out, aided by the *scavenging* action of the next charge of air entering through the intake ports. (Many two-stroke engines also exhaust through ports in the cylinder wall, rather than through the valve shown here.)

It is this ignition of fuel by the high temperature of air compressed by the piston that is the unique feature of a diesel engine, leading to to widespread use of the designation 'compression ignition engine.' however, this is an obsolete term found only in older literature; it is now accepted that the engine's original developer, Rudolph Diesel, should be honored in the name.

Compression ignition confers several advantages. Because it allows the fuel to be withheld from the cylinder until the proper point in the cycle of the piston, the engine does not experience the preignition detonation ("knock") that limits compression ratios of spark-ignition engines, and it is the permissible high compression ratio of the diesel that is the key factor in its high efficiency. Another advantage is the elimination of the need for high-volatility fuel such as gasoline. And the complex electrical ignition system of the spark-ignition engine is totally absent from diesel technology.

A unique concept of course requires unique physical features for its accomplishment; the diesel engine must have a fuel system suitable to the compression ignition principle. Other features, perhaps not unique but

12 GENERAL ENGINE PRINCIPLES

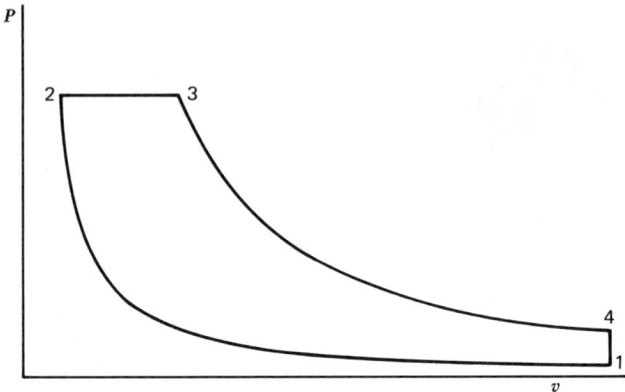

Figure 2.2 Traditional "diesel cycle" of the thermodynamics texts.

quite necessary, are also part of the complete engine. Descriptions of all of these are left to later chapters. This chapter meanwhile proceeds with exploration of the benefits and limitations inherent in a reciprocating, internal combustion, compression ignition engine.

2.1 THE DIESEL CYCLE

From a theoretician's viewpoint, the diesel engine can be looked upon as a device by which the diesel thermodynamic cycle is realized in practice. This cycle,* pictured on a pressure vs specific volume plane in Figure 2.2, consists of these four processes: reversible adiabatic compression of a working fluid (always *air* in the actual engine), its heating at constant pressure, its reversible adiabatic expansion, and its cooling at constant volume. These are, respectively, processes 1-2, 2-3, 3-4, and 4-1 in the figure. The events imposed on the working fluid by a practical engine differ somewhat from this ideal model; in particular the heating from internal combustion does not follow a constant pressure path. Indeed, there is much to be said about this cycle and its relationships to the engine it models. These explorations and amplifications shall be foregone here since the reader can easily repair to any of the many excellent thermodynamics textbooks for thorough treatments.

There are nonetheless several points about the theoretical cycle that

---

*Several different ideal cycles are used to model the diesel processes. The one used here is the traditional 'diesel cycle,' and although not the closest to the actual engine cycle, is sufficient for the development that follows here.

are of interest and of subsequent utility in a treatment such as this, and we shall have them here. Relying on your acquaintance with the thermodynamics texts for background familiarity, we note that the net work (excess of useful output over input) of the cycle is the integral of the pressure-volume relationship around the cycle, as expressed in equation (2.1) (P is pressure, v is specific volume):

$$w = \oint P dv \qquad (2.1)$$

From your acquaintance with integral calculus, you should recognize that this line integral equals the area enclosed by the cycle. And from the Theorem of the Mean, you should see that equation (2.1) can also be expressed in the form of equation (2.2):

$$w = P_m (v_2 - v_1) \qquad (2.2)$$

This equation introduces and defines the mean value of pressure P_m, a value that lies somewhere between the extreme pressures covered by the cycle. Its formal name is *mean effective pressure*, often abbreviated and symbolized MEP. Although this parameter arises from analysis of the theoretical cycle, it is more usefully applied to an actual engine. Its utility arises from a rewrite of equation (2.2) into the following form:

$$P_m = \frac{w}{(v_2 - v_1)} \qquad (2.3)$$

The specific volume difference $(v_2 - v_1)$ is a constant for a particular engine, a function of its piston stroke. Mean effective pressure consequently is directly proportional to net work, or as shall be shown in the next section, a direct measure of the power produced per unit of cylinder displacement volume.

A further point extracted from the theoretical cycle concerns the relationship of its efficiency (ratio of net work to thermal energy input) to its *compression ratio* (ratio of v_2 to v_1); high compression ratio is a major factor in high efficiency. Although so many differences exist between the theoretical cycle and the actual engine that it is fruitless to attempt a calculation of engine efficiency from the pages of the thermodynamics texts, the relationship between compression ratio and efficiency does carry over into the world of engine hardware. The analysis of the textbook cycle thus illuminates the reason for the diesel engine's high efficiency. (Recall that the introductory section has explained that compression ignition allows the engine to use high compression ratio.)

2.2 THE *PLAN* FORMULA AND THE ENGINE TORQUE AND POWER CHARACTERISTICS

The preceding section has introduced the mean effective pressure, a parameter that is proportional to the net work done by the working fluid within a cyliner upon its piston. This artificial pressure, of constant value for a given set of operating conditions, has the same effect as the actual pressure that varies continuously throughout the piston cycle. If MEP is known, it therefore gives a simple way of expressing cylinder power, as developed following:

$P \times A$ = force on piston (where A is piston area)
$P \times A \times L$ = work done on piston by the force
(L is length of piston stroke)
$P \times A \times L \times N$ = piston work rate, or *power*
(N is the number of power strokes per unit time)

The diesel power formula is therefore (rearranged from above for mnemonic purposes)

$$W = PLAN \qquad (2.4)$$

If P is in n/m^2 (or Pa), L in meters (m), A in m^2, and N in strokes/sec, then the power (W) is in watts. Alternatively, in the English system

$$W = \frac{PLAN}{33,000} \qquad (2.5)$$

Here power is in horsepower (33,000 ft-lbf/min), P is in lbf/ft², L and A in ft and ft², respectively, and N is in power strokes per minute.

Note that N is the number of power strokes per unit of time. For the two-stroke engine this is equal to the number of engine revolutions; for four-stroke it is equal to one-half of the revolutions. Since the subject of this book is an engine which is always twostroke, N is regarded hereafter as being synonymous with engine rotational speed.

The *PLAN* formula is applied either to an individual cylinder, or to a multicylinder engine. In the latter case, A must be the total area of all pistons.

The mean effective pressure is found in one of two ways, producing two slightly different values, one known as the indicated mean effective pressure (IMEP or imep), and the other as the brake mean effective pressure (BMEP or bmep).

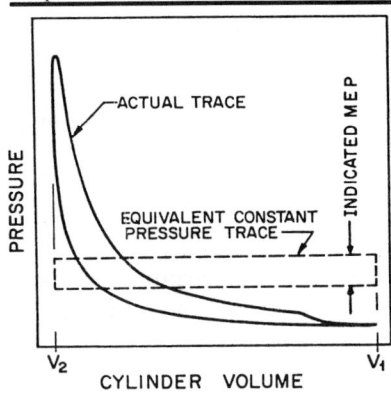

Figure 2.3 Diesel cylinder pressure–volume trace. Exhaust and intake processes of two-stroke cycle indicated by pinched area at right.

The IMEP is based on a plot of the pressure-volume relationship within the engine cylinder, the plot being made by a recording pressure gauge known as an *indicator*. The plot is integrated [equation (2.1) applied]; the resulting area (net work) divided by the volume swept by the piston stroke is the IMEP. The analytical process is the same as that applied to the theoretical diesel cycle discussed earlier. The actual cycle traced superficially resembles the theoretical, but differences are apparent. Figure 2.3 shows an actual cylinder trace; it may be compared with the theoretical cycle of Figure 2.1. Figure 2.3 also demonstrates further the concept of MEP as a measure of area enclosed by the cycle, and hence as a measure of net work.

The brake MEP is found by calculation using equation (2.4) or (2.5), following measurement of engine power by a dynamometer or a friction brake (the latter an obsolete technique that gives BMEP its name). Speed is measured also, and A and L are known from engine dimensions. With these quantities in hand, P is calculated in the obvious manner. Note that IMEP is a measure of energy applied to the engine by its working fluid; BMEP is a measure of the energy applied by the engine to an external load. BMEP is therefore always less than IMEP because the engine itself absorbs part of the energy. Internal friction is the most obvious absorber, but energy used by engine accessories (attached lube oil pumps, fuel oil pumps, etc) is also a part of the difference between IMEP and BMEP, or between their equivalents, indicated power and brake power.

The ratio of BMEP to IMEP, and of brake power to indicated power, is the *mechanical efficiency* of the engine. The difference between indicated power and brake power is often called the *friction power*.

The engine *torque characteristic* and *power characteristic* also are evident in the *PLAN* formula. Consider that power transmitted by a rotating shaft is related to torque and rotational speed by this formula:

$$W = 2\pi QN \tag{2.6}$$

If the torque Q is in newton-meters, and rotational speed N is in revolutions per second, power W is in watts. Alternatively, for Q in ft-lbf and power in English horsepower,

$$\text{HP} = \frac{2\pi QN}{33{,}000} \tag{2.7}$$

Comparison with equations (2.4) and (2.5) shows that torque is equivalent to PLA (or equal to it in SI units). The product LA is a direct function of engine dimensions, so that when a particular engine is being discussed, torque may be treated as proportional to MEP. As with power, the indicated value should be distinguished from the brake value. And the difference can be called *friction torque*.

Now reflect upon the fact that MEP is a direct measure of both torque and of net mechanical energy applied to the piston. Energy applied to the piston is very nearly proportional to the energy supplied to the cylinder by the fuel burned there. The fuel is injected in a definite amount per cylinder cycle by a positive displacement pump, a pump whose displacement is not a function of engine speed. In consequence of these relationships, torque and MEP are proportional to the amount of fuel injected per cycle (pump displacement), but not related to engine speed.

The usual "torque characteristic" of a diesel engine is a plot of torque vs engine speed (rpm). In view of the discussion above, this characteristic is a horizontal straight line, or often a family of such lines, one for each fuel setting of interest.

The corresponding power characteristic follows from equation (2.6); it is a straight line passing through the power-speed origin, and having a slope proportional to the torque.

Torque and power characteristics derived from this discussion are shown in Figure 2.4. These may be either indicated or brake torques and powers, but plots of this type most often show brake quantities.

The diesel engine is sometimes referred to as a "constant torque machine." the source of this descriptor appears in Figure 2.4—if the rpm changes without a change in fuel flow per cycle, the torque does not vary. You should realize, however, that this is an idealization. The discussion that has led to the constant torque conclusion has omitted several factors that may be significant. For one, the friction torque may be a function of speed, thus causing the brake torque to show some speed relationship also. More important in most cases is the variation in combustion air supply with speed, for the amount of fuel that will burn depends on the necessary air being present in the cylinder. The air supply of a turbocharged engine depends especially on the behavior of the turbocharger.

THE *PLAN* FORMULA AND THE ENGINE TORQUE 17

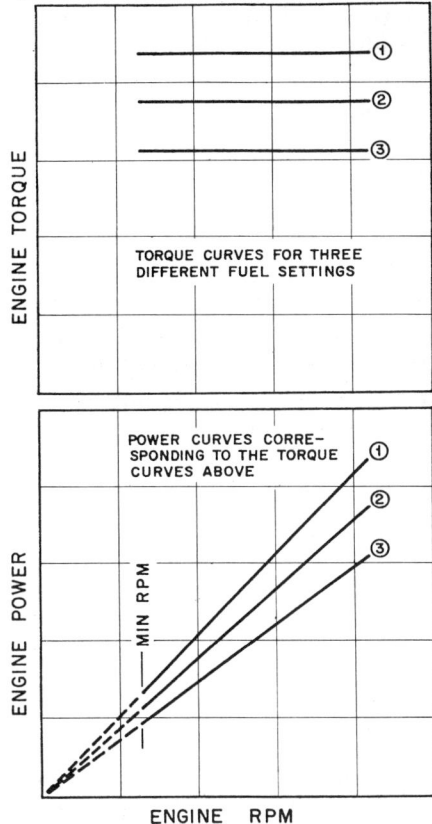

Figure 2.4 Diesel ideal torque and power characteristics.

If, for example, the turbocharger moves to an operating point of lower efficiency as engine speed declines, the resulting decline in air supply may reduce torque.

Actual torque characteristics often deviate somewhat from the ideal, especially at low engine speeds. Nonetheless, the ideal horizontal torque characteristic, and the corresponding power characteristic, are close enough to reality for use in many analyses of practical value. Chapter 5 demonstrates this in its discussion of engine characteristics matched to propeller characteristics.

In the area of ideal-vs-actual, it should be remarked that the torque referred to here, and generally throughout this book, is the average torque. Recall that the MEP is a mean value that for purposes of analysis and discussion replaces the continuously varying pressure in the cylinder. The torque which has been shown to be proportional to MEP is therefore also a steady idealization of the continuously varying torque. Since all marine engines are multicylinder, torque is smoothed to an approximate steady value by the overlapping of the individual cylinder torques, but fluctations remain. Generally they are of no consequence. An important exception occurs in the analysis of the engine as a source of torsional vibration.

2.3 PISTON SPEED

A reciprocating engine piston nominally moves in simple harmonic motion. That is, its position, velocity, and acceleration are described by a simple sine or cosine function. In practice, the motion is complicated by the rotational component of motion imposed by the crankshaft. Piston speed is thus not simple harmonic, but is expressed by

$$V_p = r\omega \sin(\omega t) + \frac{r^2\omega}{2l} \sin(2\omega t) \qquad (2.8)$$

(not exact, but accurate for $l/r >$ about 2)

where ω = rotational speed
 r = crank radius
 t = time
 l = connecting rod length

Such is piston speed, varying from negative through positive maxima in each cycle. However, when a value of this speed is listed as an engine parameter, the reference is always to an *average* speed, namely $2NL$. Since NL is a factor in the *PLAN* formula, this formula can be expressed as

$$W \times \text{MEP} \times \text{piston speed} \times A \qquad (2.9)$$

The significant thing about piston speed expressed in this convenient way is the proportionality to maximum piston acceleration force, and to the resulting bearing loads (dynamic force per unit area), as shown in the following paragraph.

If, as a reasonable approximation, simple harmonic piston speed is assumed (only first term of equation (2.8) used), the maximum value of acceleration (dV/dt) is $r\omega^2$. Note that $2r$ is the piston stroke, previously symbolized L, and that ω is proportional to N^2L, or to piston speed times N. Now, if all engine dimensions are proportional to L (a reasonably accurate assumption) then the masses of all parts are proportional to L^3, and all areas to L^2. Acceleration forces are proportional to mass × acceleration, so by the above,

F_a (piston speed $\times$ N) $\times$ L^3

F/A_a (piston speed $\times$ N) $\times$ L^3/L = (piston speed)^2

where A here is any area acted on by the force F. The forces per unit area, that is, the loadings of the bearings that must support acceleration loads, are therefore functions of average piston speed. Since permissible bearing loadings are approximately the same for all engines, piston speeds are approximately the same for all engines. A simple but important consequence is that large engines must have low rotational speeds in order to keep the NL product within the limits of bearing technology.

Table 2.1 lists piston speeds for several marine engines at their rated rotational speeds. The first two are low speed engines, the third is a medium speed two-stroke engine, and the fourth is a high speed four-stroke engine. Although the smaller engines do have higher piston speeds, the variation shown here is quite small compared to the range of engine sizes.

2.4 BASIC ENGINE DIMENSIONS

The most basic engine dimensions are *bore* (cylinder diameter) and *stroke* (distance piston moves). Directly dependent on these is the *displacement* per cylinder (also called "swept volume"), which is the circular area represented by the bore, multiplied by the stroke. Displacement per engine is therefore the product of individual displacement multiplied by the number of cylinders.

The product $A \times L$ in the *PLAN* formula is displacement, hence that formula [as most lately expressed by equation (2.9)] can be modified to

$$W \times \text{MEP} \times \text{displacement} \times \text{piston speed} \times \frac{1}{L} \qquad (2.10)$$

TABLE 2.1 ENGINE PISTON SPEEDS

Engine Type	Stroke (mm)	Speed (rpm)	Piston Speed (m/s)
Burmeister & Wain K906F	1800	110	6.6
Sulzer RND	1800	108	6.5
Electro-Motive 645	254	900	7.6
Caterpillar 3304	152	2000	10.1

20 GENERAL ENGINE PRINCIPLES

Bore and stroke appear to be independent dimensions, but in practice a relationship is evident. Table 2.2 shows the ratio of bore to stroke for a wide range of engine types and sizes.

The fundamental reason for the approximate sameness of the bore-/stroke ratios is the combined demands of compression ratio and shape of combustion space. Compression ratio has been noted earlier as the ratio of cylinder air volumes at the beginning and end of the piston stroke. Since combustion begins near the top center position of the piston, the latter volume is effectively the 'combustion space.' this space must have a reasonable ratio of its depth to its diameter (or bore) in order that fuel from the injector will be distributed uniformly throughout the air. Figure 2.5 is a cross-sectional view of an engine cylinder at the time of fuel injection, with the spray pattern shown. It illustrates quite well how the fuel distribution is matched to the space.

If, then, there is a fixed ratio r_1 of the clearance volume height s to the bore B (that is, $s = r_1 B$), and a fixed compression ratio m ($m = (L + s)/s$), then the bore/stroke ratio is

$$\frac{B}{L} = \frac{1}{r_1(m-1)} \qquad (2.11)$$

Equation (2.11) shows that bore/stroke ratio is fixed if compression ratio and shape ratio of clearance volume are fixed. Since the latter two parameters must be approximately the same for all engines, the bore/stroke ratio is likewise nearly the same.

2.5 MORE ON MEAN EFFECTIVE PRESSURE

Engines are heavy; their weight subtracts from vessel payload. Moreover, the materials of which they are built are expensive. In consequence, it is important that the output per unit of engine size be as high as possible. A glance at equation (2.10) then shows that the product of MEP, piston

TABLE 2.2 ENGINE BORE/STROKE RATIOS

Engine Type	Bore (mm)	Stroke (mm)	Ratio
Burmeister & Wain K906F	906	1800	1.99
Sulzer RND105	1050	1800	1.71
Electro-Motive 645	230	254	1.10
Caterpillar 3304	121	152	1.26

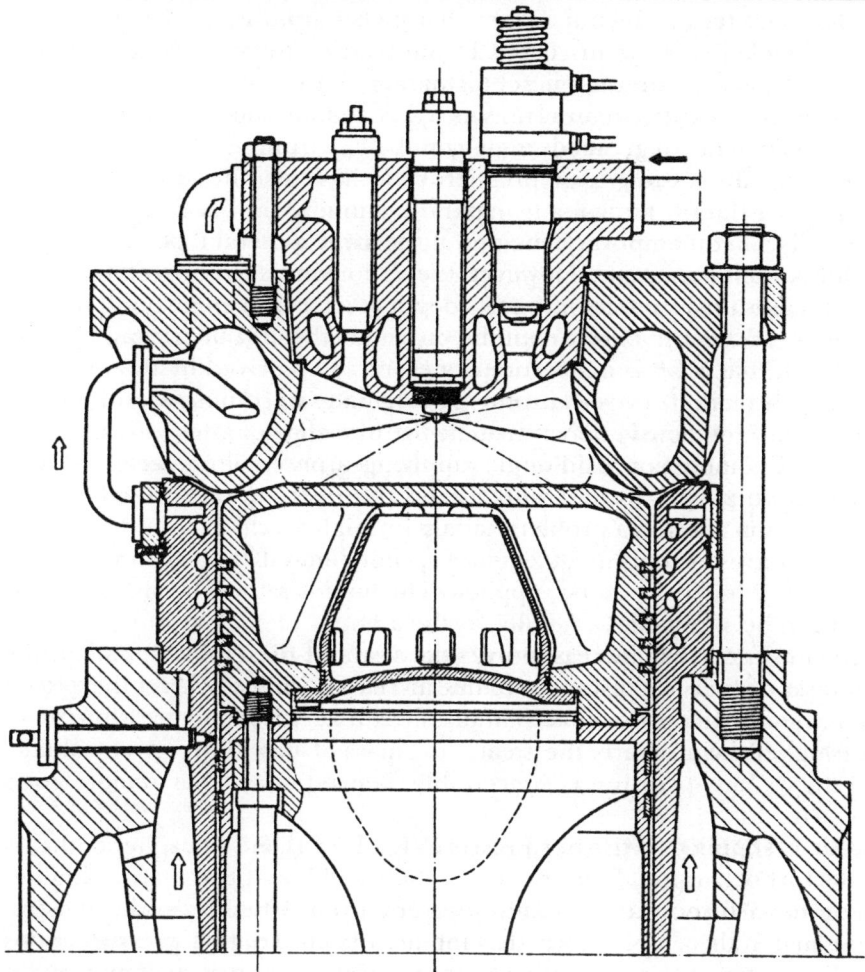

Figure 2.5 Cross section of engine cylinder showing spray from the injector. Sulzer RND engine.

speed, and $1/L$ should be as high as possible. The continued development of the diesel engine in competition with steam, and often among its own types of engines, is therefore a three-pronged push to advance the technology in each of these areas.

Piston speed has been discussed in Section 2.3, with the general message that this parameter is limited by the upper bounds of feasible bearing loadings. Of course, this bound is not entirely static; an engine designer can reduce the masses of reciprocating parts, enlarge bearing areas, and seek bearing materials that can carry high loadings. Each of these in turn has its limits.

22 GENERAL ENGINE PRINCIPLES

The $1/L$ term, which indicates that high output is to be gotten from short stroke, is to be discussed in the next section. Its general consequence is a branching of diesel technology into distinct engine types. A seeking after its advantages leads into the medium speed field. Within the low speed field, no great advantage can be had from short stroke. Indeed, there is some incentive for long stroke (as will also be discussed later).

The MEP term remains. It, naturally enough, has its limits also, and typically all contemporary engines of the same general type have similar values of this parameter. However, the opportunities to expand the limits are greater than for piston speed, so that advances in engine output have been—and apparently will continue to be—in the area of increasing MEP.

Recall that MEP is a measure of net energy received by the piston as it goes through its cycle. The energy originates in combustion of fuel, an occurrence obviously dependent on the presence of adequate air at the point of combustion. Although supplying a precisely metered mass of fuel at the proper instant with the proper distribution throughout the combustion space is a problem calling for high mechanical and hydraulic technology, the output of engines is not limited by any limit on the amount of fuel that can be supplied. The limit is rather the amount of air that can be supplied. Generally, as the amount of air that can be packed into a cylinder has increased over the years of diesel development, the amount of fuel per piston cycle has increased similarly. And the reason that the burden rests on the air supply is that it is a compressible fluid; the higher the pressure, the greater the mass of air within a given volume. In theory, any pressure is conceivable, hence practice has remote goals to strive for.

Air is supplied at greater pressure—greater than atmospheric, that is—by turbocharging. Since the early 1960s, all low speed marine engines have used turbocharging. Development toward higher mean effective pressure is therefore largely development toward higher pressure ratios (ratio to atmosphere) supplied by these exhaust gas driven compressors.

Increases in fuel burned through increases in air supply have a negative consequence that may provide the ultimate limit on the feasible amounts of the two fluids: the higher thermal and mechanical stresses to engine structure.

Mechanical stresses are a consequence of the pressure within the cylinder, and are, naturally, higher when MEP is higher. On the other hand, this mean pressure is more truly a difference between pressure applied to the piston in its power stroke, and pressure applied by the piston in its compression stroke. If compression ratio (hence compression pressure) is lowered, a given maximum pressure produces a greater MEP. Thus it is that compression ratios have been lowered as pressures developed by turbochargers have become higher; typical values would be about 20 for earlier naturally aspirated (nonturbocharged) engines, to about 16 for the turbocharged engines of the 1970s. But lower bounds on compression ratio restrict further reduction. The requirements of compression ignition form an obvious and important bound.

Mechanical stresses can also be reduced by greater structural cross sections to transmit the loads. This approach runs counter to the objective of greater output per unit of engine size and weight, and also worsens thermal stresses. Ths latter category of stresses is a consequence of nonlinear temperature gradients in the metal boundaries of the combustion space, a situation exacerbated by greater thicknesses. The developmental process is therefore toward more skillful design of load-carrying parts, and of the heat flow paths from sources of high temperature to the engine cooling system.

The flow of heat is a problem of special consequence for the low speed engine because its large pistons represent a long path for the heat absorbed near their centers. Without some means of cooling applied directly to the piston, this heat must flow by conduction to the outer circumference of the piston, through the piston rings into the cylinder liner, and thence into water circulating through cooling passages. The larger the piston diameter, the longer the path, and consequently the higher will be the temperature near the center of the piston face. The remedy is to cool the piston, most commonly with water introduced into internal cooling passages via telescoping pipes.

In summary we note that the goal of higher mean effective pressure—this to increase the output per unit of engine size and weight—is sought by compressing more combustion-supporting air into the cylinder, accompanied by measures that enable the boundaries of the combustion space to withstand the consequent mechanical and thermal loading. The latter factor is usually the limiting one.

2.6 STROKE LENGTH

Equation (2.10) indicates that engine power per unit of displacement is inversely proportional to stroke length. Shortening stroke is therefore a route to greater output from a given size of engine, but not one that is free of limiting factors. Displacement is proportional to stroke length (displacement $= A \times L$), so that a decrease in stroke requires an increase in piston area. The limiting factor here is bore/stroke ratio; the requirement for approximately constant bore/stroke demands that a decrease in stroke be accompanied by a decrease in piston area.

Since bore/stroke ratio is restricted only by the configuration of the individual cylinder, the freedom to divide total piston area among a number of cylinders is an expedient by which the benefits of short stroke can be had without upsetting the ratio. This, perhaps more than any other factor, is the reason that practical engines always consist not of one large cylinder (which would doubtless be the simplest engine), but of many smaller cylinders. But limits exist in the direction of numerous cylinders also. Among low speed engines, it's a matter of crankshaft flexibility when this component must be long to accommodate many cylinders, and perhaps also of practicable length for a ship machinery space. Twelve cylinders are the greatest number offered by the low speed builders.

Stroke length is also a factor in piston speed; hence a decrease in stroke allows an increase in engine speed. The requirements of ship propeller efficiency deny this route to greater output for an engine that is to be connected to the propeller without speed reduction (propeller efficiency usually demands *lower* speed). If one accepts the suitability of a speed reducer, then a radical increase in speed is possible when stroke is reduced and number of cylinders increased. Thus the medium speed ship propulsion engine, with speeds as high as 900 rpm, and thus the remark in the preceding section on 'branching of diesel technology.' The branching is into low speed and medium speed philosophies, with the former remaining faithful to direct connection, and the latter seeking (and finding) greater output from a given engine size at the cost of greater complexity (more cylinders, the reduction gear), and of slightly lower (usually) engine efficiency.

An *increase* in stroke length benefits a direct connected propulsion engine because it permits a decrease in propeller speed, with consequent improvement in propulsive coefficient. The disadvantages of this measure are the increase in cylinder displacement and the larger crankshaft (reflect that stroke length must be proportional to crank radius). The benefit side has sufficient appeal that some engine builders take advantage of progress toward higher MEP to lengthen stroke (trading MEP vs $1/L$ in equation (2.10)) rather than to increase engine power.

2.7 SUMMARY OF ENGINE LIMITATIONS

This chapter has outlined the factors that determine the general configuration and characteristics of a diesel engine. The heart of the matter is in the simple *PLAN* formula, first seen as equation (2.4), and last as equation (2.10). It illustrates that engine power output is a product of its mean effective pressure, its displacement, and its rotational speed (2.4), or alternatively, as a product of mean effective pressure, piston speed, and piston area (2.9), or in still another way as a product of mean effective pressure, displacement, piston speed, and the inverse of stroke length (2.10).

Average piston speed is proportional to the product of rotational speed and stroke length. Because it is also proportional to accelerative forces per unit area, piston speed is effectively a constant, so that rotational speed and stroke length are inversely proportional.

Stroke length is related to cylinder bore, and hence to displacement per cylinder, by the requirements of bore/stroke ratio. Reasonable limits for height-to-diameter proportions of the combustion chamber (effectively, the clearance volume above the piston at its top center position) fix this ratio within a narrow range. The advantages of short stroke can thus be had only through the expedient of multiple cylinders, an option that leads to medium speed engines as an alternative to low speed.

High output per unit of engine size and weight—tantamount to high output per unit of displacement—is greatly to the competitive advantage of any engine. Given the several restrictions previously mentioned, the principal route open to the low speed diesel is increase in mean effective pressure. The twin requirements here are greater mass of combustion air per unit of displacement, and a parallel ability of the engine to support the resulting mechanical and thermal stresses.

2.8   NOTATION FOR CHAPTER 2

A	piston face area
A	bearing area
B	cylinder bore
BMEP	brake mean effective pressure
bmep	brake mean effective pressure
ft lbf	pounds force
HP	horsepower
IMEP	indicated mean effective pressure
imep	indicated mean effective pressure
kPa	kilopascal
L	piston stroke length
l	connecting rod length
MEP	mean effective pressure
mep	mean effective pressure
m	compression ratio
m/s	meters per second
N	rotational speed
N/m^2	newtons per square meter
Q	torque
P	mean effective pressure
P_m	mean effective pressure
r	crank radius

r_1	ratio of s to B
rpm	rotational speed, per minute
s	height of cylinder clearance volume
SI	international system of units
t	time
V_p	instantaneous piston speed
v	specific volume
W	power in watts or kilowatts
w	net work
ω	rotational speed (radian measure)

Chapter Three

ENGINE CONSTRUCTION

Chapters 1 and 2 have introduced the main features and behaviors of the diesel engine, with due emphasis on the low speed marine type. Figures 1.4 and 1.5 should have given you a conception of its general configuration, as well as a few details if you observed closely. This chapter proceeds further into the details of the low speed engine and into the functions and appearance of its parts, and mentions some of their design considerations.

An excellent source for familiarity with marine diesels of all types is the magazine *THE MOTOR SHIP* (monthly, London). Among other things, it occasionally publishes wonderfully detailed cutaway views of marine engines, with numerical keys to indicate the engine parts. Figure 3.1 is a reproduction of one such illustration, that of a Burmeister & Wain (B&W) K90GF engine. It serves this chapter as a pictorial introduction and a reference point for the illustrations of individual parts that are to follow. Some of the details of rating, sizes, and weights for this engine are given in Table 3.1.

3.1 PISTONS AND RUNNING GEAR

We can scarcely claim that one part of an engine is more important than another, but certainly the piston is the heart of the engine—of any recip-

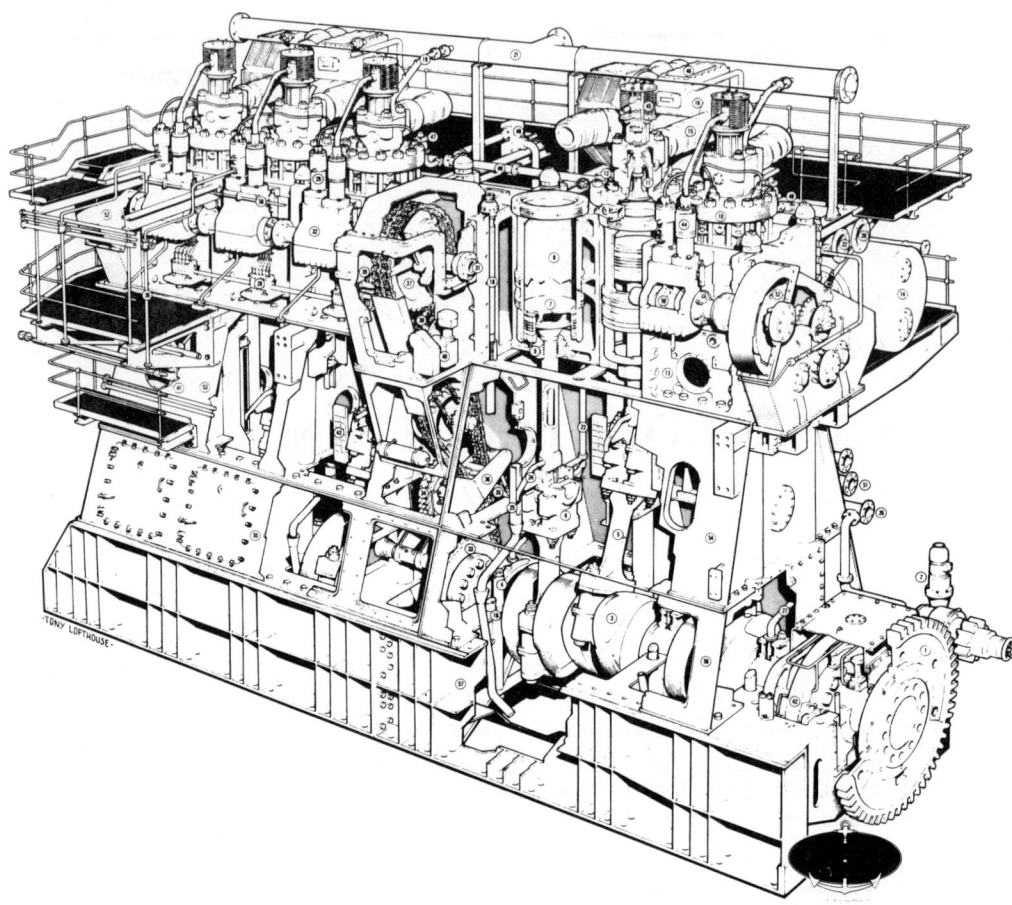

Figure 3.1 Burmeister & Wain K90GF Engine (reprinted, by permission, from *THE MOTOR SHIP* journal).

1 Turning wheel
2 Turning gear motor
3 Crankshaft web
4 Main bearing
5 Connecting rod
6 Crosshead assembly
7 Piston assembly
8 Cylinder liner
9 Piston rod gland
10 Cylinder head assembly
11 Exhaust valve
12 Fuel valve (3)
13 Cylinder block/scavenge air receiver
14 Air inlet manifold
15 Cylinder exhaust manifold
16 Turbocharger
17 Air cooler
18 Tie bolt
19 Cooling water outlet branch pipe
20 Cooling water inlet main
21 Cooling water returns main
22 Cooling oil supply main
23 Piston cooling telescopic pipes
24 Piston coolant branch pipes
25 Piston cooling outlet collector
26 Lube oil supply main
27 Lube oil branch pipes

TABLE 3.1 PARTICULARS OF THE ENGINE SHOWN IN FIGURE 3.1

Detail	B&W 6K90GF
Number of cylinders	6
Cylinder bore	900 mm
Piston stroke	1800 mm
Continuous service rating	3100 hp/cylinder
Speed	110 rpm
Mean effective pressure (indicated)	11.1 kg/cm^2
Overall length	1393 mm
Width of mounting	440 mm
Height above shaft	943 mm
Depth below shaft	175 mm
Overhauling height above shaft	1150 mm
Weight	
Total	645 t
Cylinder liner	3.5 t
Piston, with piston rod	3.0 t
Connecting rod	2.5 t
Crosshead	2.6 t
Cylinder cover	5.3 t
Crankshaft	145 t
Turbocharger rotor	0.6 t
Largest part handled (overhaul)	5.3 t

28 Cylinder lubricator pumps
29 Fuel pump
30 Fuel supply line
31 Camshaft
32 Cam boxes
33 Crankshaft chain drive sprocket
34 Chain tensioning jockey wheel
35 Chain adjustment
36 Chain tensioning bell crank
37 Reversing gear and camshaft drive
38 Governor and air distributor drive gear
39 Air distributor
40 Woodward governor
41 Lifting beam
42 Thrust block
43 Crosshead slipper
44 Valve actuating pump
45 Valve hydraulic assembly
46 Turbocharger exhaust outlet
47 Air starting valve
48 Air start master valve
49 Air start manifold
50 Fuel and valve pumps drive assembly
51 Air cooler water supply lines
52 Moment compensators
53 Upper frame unit forward
54 Upper frame unit aft
55 Lower frame unit forward
56 Lower frame unit aft
57 Bedplate
58 Fuel pump control shaft

30 ENGINE CONSTRUCTION

rocating engine—for it is the component that interacts with the working fluid to perform the key step of converting thermal energy to mechanical energy.

The piston is often considered to be a part of the *running gear*, consisting of piston assembly, piston rod, crosshead, and connecting rod. Figure 3.2 illustrates a typical running gear, this one from the Sulzer RNDM engine.

The piston is always an assembly (hence the term above) of several parts, the principal ones being a top or "crown," a skirt, and an inner structural part of one or more pieces that supports the other two and

Figure 3.2 Running gear (piston, piston Rod, crosshead, connecting rod) for Sulzer engine (Sulzer photograph).

PISTONS AND RUNNING GEAR 31

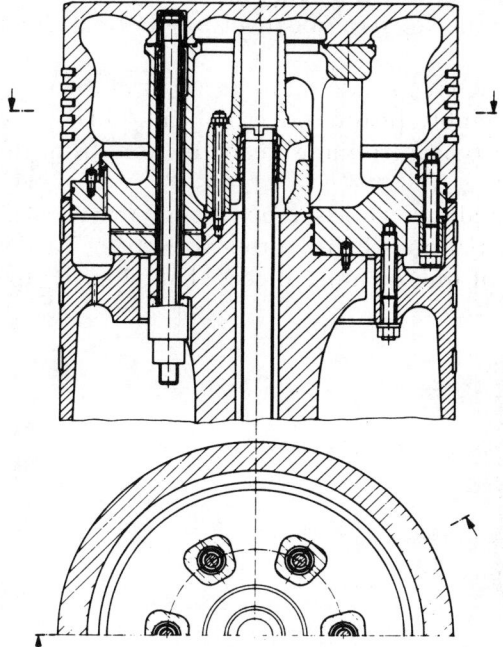

Figure 3.3 Piston cross section (MAN drawing).

connects them to the piston rod. The parts can be distinguished in Figure 3.3, a cross sectional view of a piston.

The top or crown is subject to heavy loads of both mechanical and thermal nature, arising from the high pressure and high temperature, respectively, of the cylinder gas. Thick metal cross-sections, the common design response to high mechanical loads, are unwise here because they increase temperature differences and hence increase stresses caused by the thermal loads. As a consequence, this part is always of rather thin-walled construction, but well stiffened by integral ribs and extensive support from the piston structural parts. The top is either a casting or forging of steel, either carbon steel or a heat-resistant molybdenum alloy.

The lands (grooves) for the piston rings are located around the circumference of the piston top. Because the rings move slightly with respect to the piston, some wear is experienced by the lands; to minimize wear the land surfaces are usually hardened and ground, and may be chrome plated.

The skirt forms the lower periphery of the piston. Its function is to cover the ports in the cylinder wall while the piston is near the top of its

32   ENGINE CONSTRUCTION

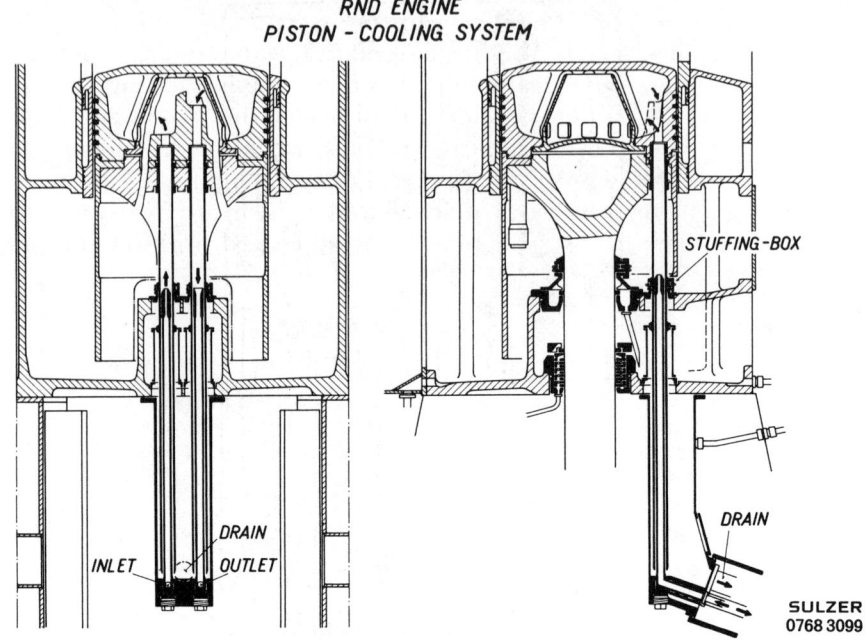

Figure 3.4   Piston cross sections, showing the telescoping pipes for water coolant supply and drain (Sulzer drawing).

cycle to prevent a short circuit between inlet and exhaust during this time. On the other hand, uniflow engines, which exhaust through valves in the cylinder cover, do not require this function and thus do not require the piston skirt.*

As in all reciprocating engines, the piston rings perform the essential service of making the seal between piston and cylinder wall. In spite of their rather simple nature, their burdens of high pressure differences, heavy sliding forces, and deposits from poor fuel or poor combustion require that they be designed and manufactured with care.

Skirts are usually fitted with "scuffing bands," these being rings of a comparatively soft material such as a lead bronze. They bear the sliding contact between piston skirt and cylinder wall.

Cooling of the piston, especially the top, is essential in the low speed

---

*Chapter 6 will explain the functions of the ports and the several types of scavenging such as the uniflow just mentioned.

engine. Either water or lubricating oil is used as the medium. Conveying either fluid to and from the moving piston is, naturally, something of a problem, a problem that is a bit more significant with water since leakage of that fluid within the engine can contaminate the lubricating oil. The typical solution with water is a sliding joint (called the "trombone" after the resemblance to that musical instrument), well sealed, between stationary pipes, with the pipes attached to the piston. Figure 3.4 illustrates the arrangement, and the pipes attached to the piston are clearly evident in Figure 3.2. The piston of Figure 3.3 is oil cooled, with oil entering

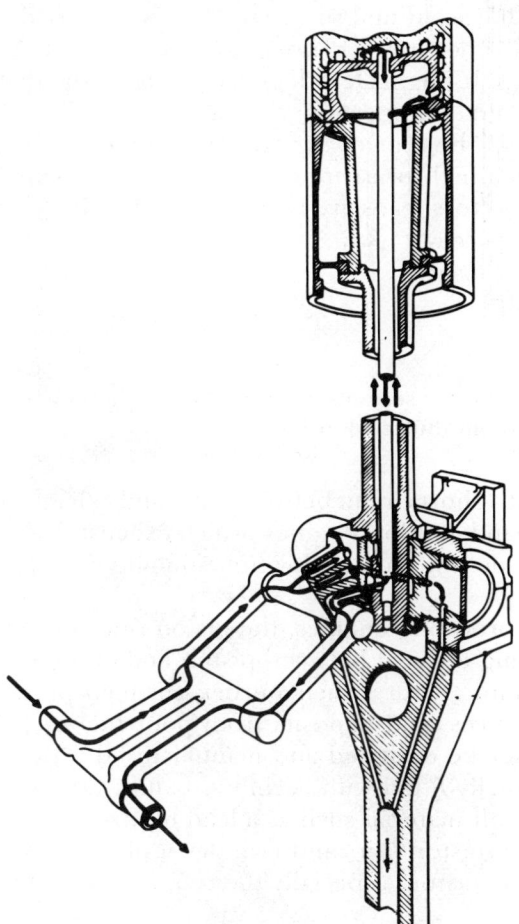

Figure 3.5 Cutaway view of engine running gear, showing swing pipes for coolant supply and drain.

34 ENGINE CONSTRUCTION

through a centerline pipe evident in that figure. Figure 3.5 shows the route of the oil in and out via the crosshead and piston rod, with swinging joints making the connection between the stationary lube oil piping (extreme left in the figure) and the moving crosshead.

Other elements of particular concern in the running gear are its bearings: the crankpin, crosshead pin, and crosshead guide bearings. These bearings are generally of a common bearing metal ("white metal"), backed by a steel shell. The bearing for the crosshead pin (pin that connects crosshead to connecting rod) is the one receiving greatest design care. Its service is severe because it does not allow buildup of the hydrodynamic lubricant film so essential to heavily loaded bearings. Reflect that the connecting rod does not rotate about this bearing, but merely oscillates through a few degrees, and slowly at that, given the low speed of the engine in question; a hydrodynamic film is unlikely under such conditions. The remedy is hydrostatic lubrication, that is, oil supplied under such high pressure that the crankpin floats hydrostatically. To this end, the crossheads are typically supplied by their own lubricant

Figure 3.6 Crankshaft for Sulzer RNMD engine (Sulzer photograph).

pumps, since this pressure is higher than that required generally by the lubricating system.

3.2 CRANKSHAFT

Figure 3.6 shows the crankshaft for a Sulzer six-cylinder RNDM engine, and Figure 3.7 shows the same shaft mounted in its bearings during assembly of the engine. The smaller toothed wheel drives engine accessories, in particular the camshaft that operates fuel pumps, lubricators (for example, the small pumps that inject lubricant into the cylinders), etc. The larger toothed wheel is the flywheel; its teeth mate with a turning gear that can rotate engine, shaft, and propeller at low speed. The thrust collar is visible just behind the flywheel.

Figure 3.7 Crankshaft for Sulzer RNMD engine mounted in bedplate (Sulzer photograph).

36 ENGINE CONSTRUCTION

The crankshaft is an assembly of steel forgings, or of steel castings and forgings, joined by shrink fitting. If all webs and pins are made individually, the shaft is referred to as a "built" shaft. If each cylinder set, that is, a pair of webs with the crankpin between them, is made as a unit and joined to its neighbor by the shrunk-in shaft journals, the shaft is "semi-built." the third alternative—a single forging—is unusual in low speed engines. In addition, crankshafts for engines of more than six cylinders may be constructed from two lengths bolted together.

Figure 3.8 Cylinders being assembled for Sulzer RNMD Engine. Liner Is being fitted into nearest cylinder (Sulzer photograph).

3.3 CYLINDER BLOCK AND CYLINDER LINER

The familiar automotive engine (and high speed marine diesels as well) are built on the "monoblock" principle; their major static component is a single piece cast "block" into which are cast and bored the individual cylinders. In contrast, the low speed diesel cylinder is an independent unit. From three to twelve identical units are made individually and bolted together to form the central structural element of the engine.

Each of these cylinders consists of two major parts: the block or structural member and the liner. Figure 3.8 shows cylinders being assembled for a Sulzer RNDM engine, with the liner being lowered into one of the cylinder blocks. Exhaust ports (the row of near-square holes) and scavenge ports (air inlets, below the exhaust ports) are quite evident in the liner. On the block one may note the surfaces by which it mates to the next cylinder. Observe, for example, the two semicircular channels which each form half of a bore for the long tie bolts that hold the engine

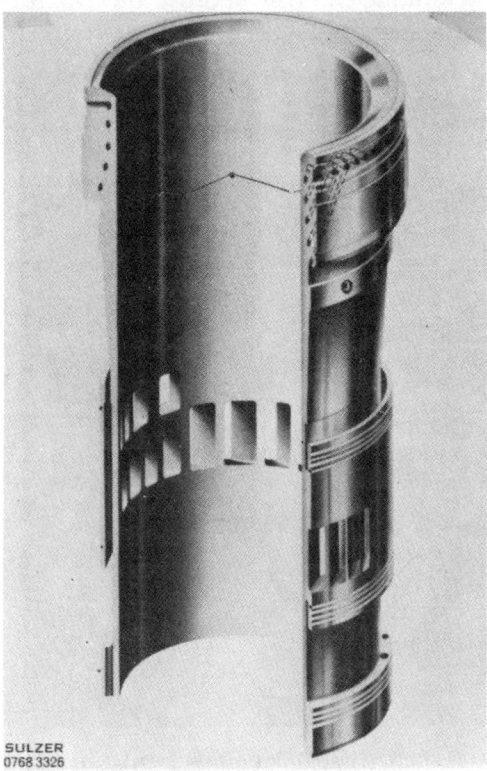

Figure 3.9 Cylinder liner for Sulzer engine (Sulzer photograph).

38 ENGINE CONSTRUCTION

together vertically. Figure 3.9 is a cutaway view of a cylinder liner, again showing the exhaust and scavenge ports. Other points to note are the small ports and grooves in the upper wall by which cylinder lubricant is distributed, and the holes near the top (exposed by the cut) by which cooling water is circulated through the liner. These coolant passages are in addition to those around the outer surface of the liner, above and below the band of ports. Their presence indicates "bore cooling," so-called after these small bores.

The liner bears continuous heavy sliding contact from the piston rings, and is attacked by the acid products of combustion. The cylinder lubricant resists both of these burdens, and the liner is made of cast iron alloyed for wear resistance. Nonetheless, the liner experiences significant wear. "Significant" here implies a degree sufficient to degrade the performance of the engine below acceptable limits within a time that is a small fraction of the life of the ship. It is therefore regarded as a replaceable part and, naturally enough, is designed for easy removal.

Figure 3.10 is a half-sectional view of a cylinder liner for a Doxford engine. This is of unique interest because the Doxford is an opposed-piston engine and hence the liner shown here represents, in effect, two cylinders.

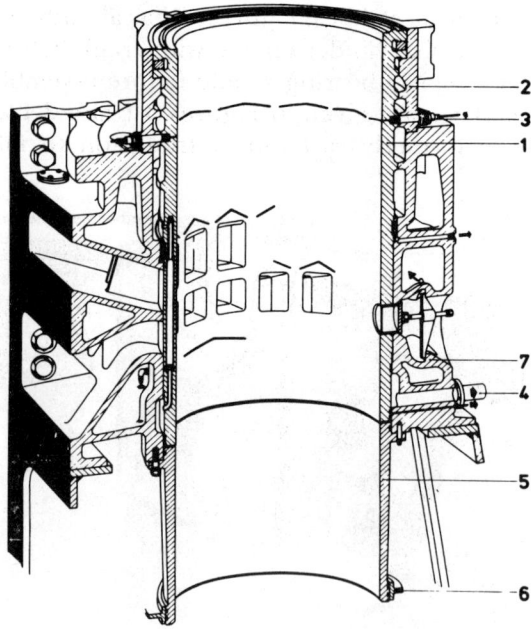

Figure 3.10   Cylinder liner for an opposed-piston engine (MAN drawing).

GENERAL STRUCTURE   39

3.4 CYLINDER COVERS

The cylinder cover or "head" is subjected to the same pressure and thermal loads as the piston, although being a stationary part the design solutions are arrived at more easily. The thin-walled, stiffened and cooled construction is often used here also. The cover is a mounting for accessories such as fuel injectors, air starting valves, cylinder relief valves, and exhaust valves (if used). The configuration varies considerably with the type of engine. For example, only the B&W engine uses a valved exhaust, and to go to the extreme, an opposed-piston engine such as the Doxford requires no cover at all.

Cylinder covers appear in several figures. For example, Figure 3.18 shows an external view, and Figures 1.4, 2.5, 6.14, and 8.5 show cross-sections.

3.5 GENERAL STRUCTURE

A rigid seating for the crankshaft is essential, and this is provided by a "bedplate" (in quotes here because it is scarcely a plate, as our illustrations will show you) constructed of deep longitudinal beams and transverse crankshaft bearing saddles. Figure 3.8 has already shown a top view of the bedplate for a six-cylinder engine with crankshaft in place. Figure 3.11 is a view of a typical bearing saddle before assembly. Other views of complete bedplates are given in Figures 3.1 and 3.14.

Bedplates were formerly cast, but a transition to fabrication from

Figure 3.11 Crankshaft bearing saddle (Sulzer photograph).

welded steel plate was made in the 1950s. This change allowed easier construction by the engine builders' shipyard licensees who lacked the necessary foundry capacity for such large castings, and had the additional benefits of reducing engine weight and of producing a stiffer bedplate, the latter being mainly due to the higher modulus of elasticity of steel vis-a-vis cast iron.

The cylinders are supported from the bedplate by a boxlike frame, typically constructed from longitudinal plates over transverse "A-frames" (from the usual shape). Figure 3.12 shows an A-frame being fabricated, and Figure 3.13 shows an assembly.

Figure 3.14 is an exploded view of an overall structure showing the bedplate, cylinder block, and the supporting structure just mentioned. The last is, in this case, constructed of an upper and lower half.

Figures 3.15, 3.16, 3.17, and 3.18 are a set of photographs of a MAN engine being assembled in the shop. The three lower parts in Figure 3.14 are shown again by Figures 3.15, 3.16, and 3.17, respectively. Figure 3.18 shows the top of the engine with cylinder covers in place.

The essential ingredient tying the engine together in the vertical direction is the set of tie rods, a pair on either side of each cylinder, running from bedplate through the cylinder blocks. The rods are shown but poorly in Figure 3.1, and not at all in subsequent figures, but the holes and channels for them are clearly evident in several of these figures, giving good indications of their location and size. These rods are usually prestressed (by torquing their nuts) sufficiently to keep the structure between bedplate and cylinder cover under compression.

3.6 CYLINDER PORTS AND VALVES

The two-stroke engine requires no valves for admission and discharge of its working fluid, these functions being readily accomplished via ports in the cylinder wall that are covered and uncovered by the piston. These ports show in several of the previous illustrations, for example, Figures 3.8, 3.9, and 3.10. As an alternative, exhausting via a valve in the cylinder cover allows timing of the exhaust process to be controlled, and brings the advantages of uniflow scavenging.* The majority of the low speed engine builders favor the ported exhaust; only Burmeister & Wain and Mitsubishi use the valved exhaust. The engine of Figure 3.1 is B&W, and the exhaust valve and its opening gear can be seen there. (Look at the right-most of the two cylinders that are shown in cutaway views.) The

---

*The several kinds of scavenging are discussed in Chapter 6.

Figure 3.12 Engine A-frame (GMT photograph).

Figure 3.13 Engine frame during assembly (Sulzer photograph).

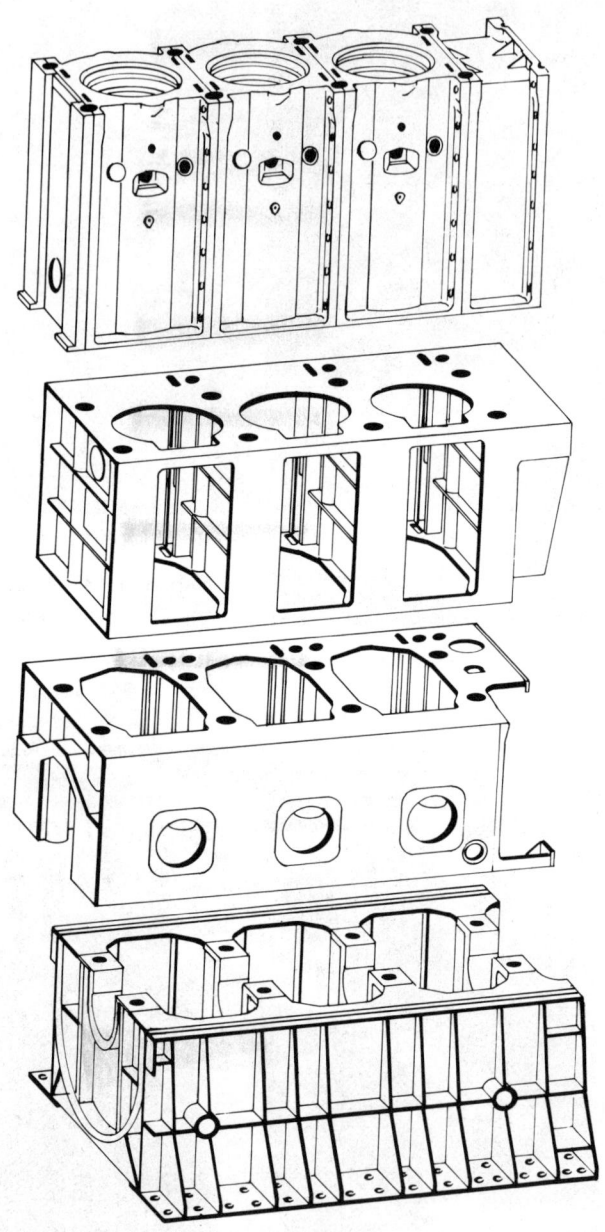

Figure 3.14 Exploded assembly view, with three cylinders shown (MAN drawing).

Figure 3.15 Engine under assembly (MAN photograph).

Figure 3.16 Engine under assembly, continued (MAN photograph).

Figure 3.17 Engine under assembly, continued (MAN photograph).

Figure 3.18 Engine under assembly, continued (MAN photograph).

FUEL PUMPS AND INJECTORS 45

B&W exhaust valves are hydraulically actuated. Each cylinder is equipped with a hydraulic pump driven by the camshaft that lies at cylinder level along the near side of the engine in Figure 3.1. The pipe leading from pump to valve actuator is clearly evident at each cylinder in this figure.

B&W has also used mechanical valve actuation. With this alternative, push rods and rocker arms are prominent features of the engine in the vicinity of the cylinder covers.

Low speed engines are started by compressed air admitted to some of the cylinders. At least three cylinders must be equipped for air admission to ensure that the engine can start from any crank angle; those cylinders so equipped have a "starting air valve." one of these valves shows (though poorly) in Figure 3.1, and a sketch of a valve and the air distributor that actuates it via a pilot valve is given in Figure 3.19.

Cylinder relief valves, set to open at some designated pressure above the maximum firing pressure in the cylinder, are required by the classification societies (for example, the American Bureau of Shipping).

A rather specialized cylinder valve is the gas admission valve, found only on engines equipped for burning gaseous fuel in addition to oil. Chapter 8 includes discussion of gaseous fuel, and the valve can be seen in Figure 8.5.

3.7 FUEL PUMPS AND INJECTORS

That fraction of the fuel system which is directly a part of the engine* consists of the fuel pumps and the injectors.

Each cylinder is typically provided with its own fuel pump, driven by the engine camshaft. The pumps are visible in Figure 3.1, paired with the hydraulic exhaust valve pumps previously mentioned. The cross-section of such a pump appears in Figure 3.20. The cam, camshaft, and cam follower roller are evident at the lower part of the figure. The plunger, which is moved by the cam follower and which accomplishes the pumping action, lies along the centerline in the top half of the unit. The effective displacement of the plunger, and hence the amount of fuel pumped on each stroke, depends upon its rotational alignment within the barrel of the pump, since this alignment determines when the fuel inlet port is covered by the plunger. It is rotated to achieve the desired position by a shaft from the engine governor. The control shaft is visible in Figure 3.1.

---

*That is, excluding heater, purifiers, tanks, etc, which are not part of the engine, and therefore are discussed in chapter 10.

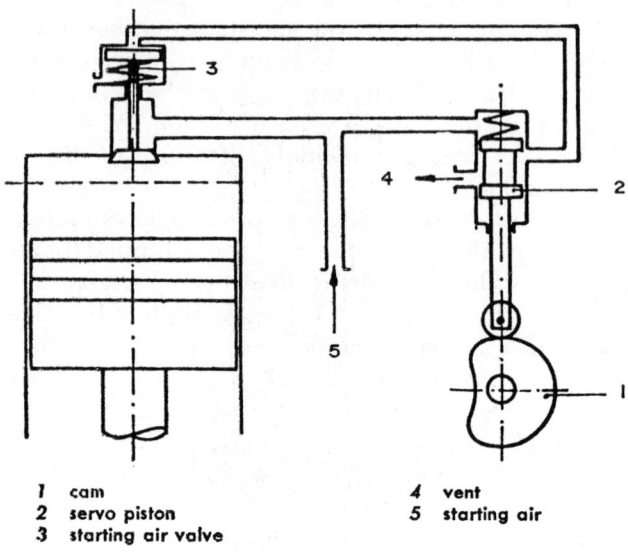

1 cam
2 servo piston
3 starting air valve
4 vent
5 starting air

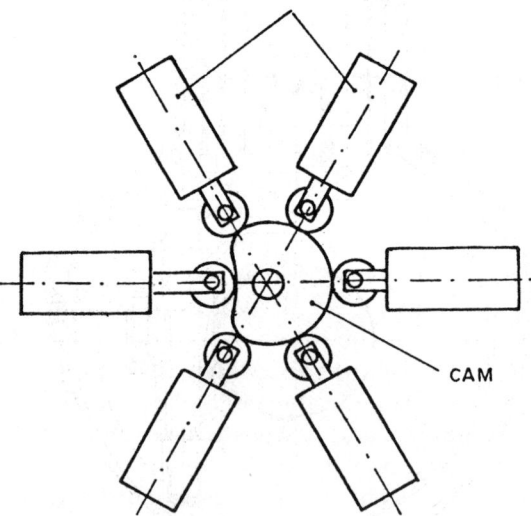

Figure 3.19 Starting air valves. Valves for one cylinder at top. All distributing valves for engine at bottom (from *Marine Engineering;* by permission of Society of Naval Architects and Marine Engineers).

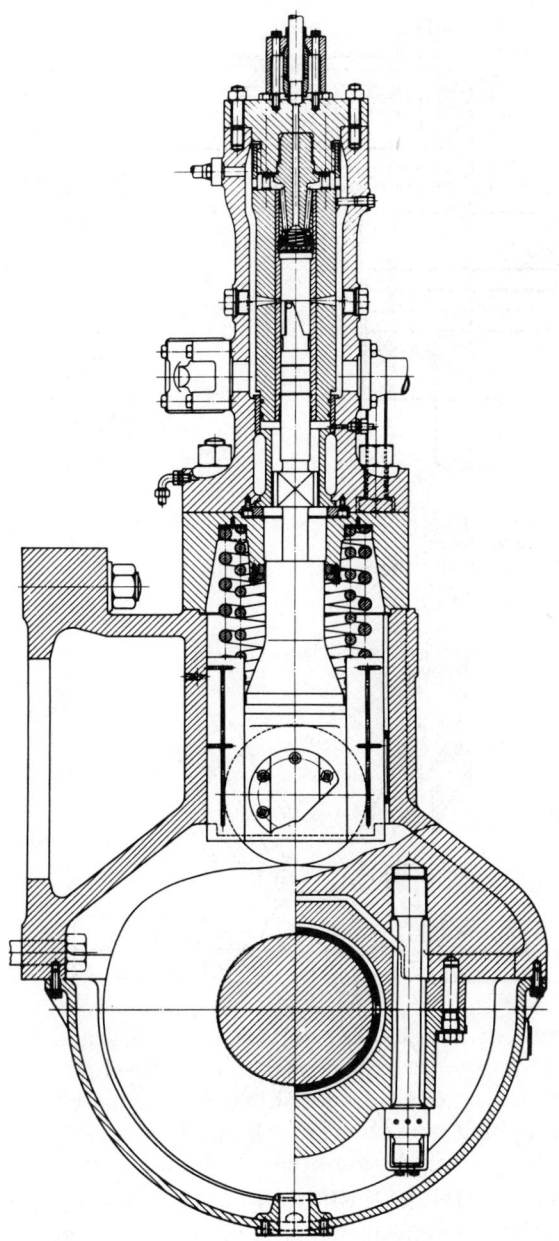

Figure 3.20 Cross section of fuel pump.

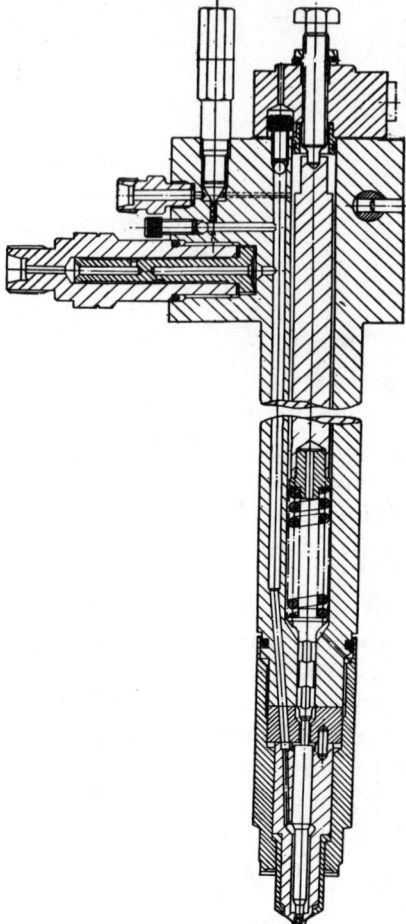

Figure 3.21 Cross section of injector.

The fuel enters the engine via the *injector*, which is a combined valve and atomizer. The valve is a spring-loaded needle check valve that opens when the pressure reaches a sufficient level, usually about 1/3 of its maximum level (which is in the range 50 to 100 MPa). Atomization into the cylinder occurs through fine holes at the injector tip. Figure 3.21 is a cross-section of a typical injector, and Figure 3.22 is a photograph of an injector tip, partly cut away to show the valve seat and coolant passages around the tip. The spray passages are visible in this figure, though difficult to see because of their fineness. Figure 3.23 is a typical map of fuel pressure, and of fuel pump stroke and velocity as a function of crank angle.

AIR AND EXHAUST   49

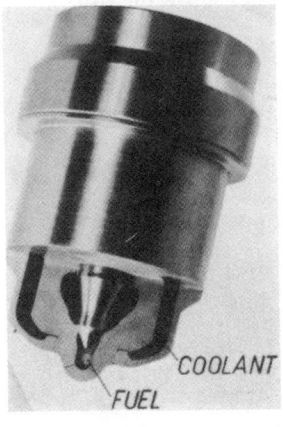

Figure 3.22 Injector tip (Sulzer photograph).

Microprocessors began to appear as engine control components in the late 1970s. Their use to control fuel injection allows major changes in the system described in the preceding paragraphs. Figure 3.24 shows the essential features of "electronic fuel injection" as introduced by MAN circa 1980. In this system, two or more fuel pumps, driven by gears and cams from the crankshaft, pressurize a fuel accumulator that maintains injection pressure at about 70 MPa. A hydraulically actuated valve at each cylinder, controlled by the fuel control microprocessor (part of the "elektronic-hydraulic unit" in the figure), admits fuel with the proper timing to each injector. The camshaft that actuates the fuel injector in the more traditional system described previously is therefore not required in an engine with the electronic injection method.

3.8 AIR AND EXHAUST

Air must be supplied to a two-stroke engine to *scavenge* its cylinders of spent cylinder gas; the air that is trapped at the end of this process becomes the working fluid for the next cycle. Since the 1960s; all low speed engines have been built as turbocharged machines, hence the principal element in supplying the air is the exhaust gas driven turbocharger. However, many engines combine this unit with a reciprocating air pump, such as the undersides of the pistons themselves. And electrically driven blowers to supply air at low engine loads are typical. There are several combinations, several methods of scavenging, and several methods of using the exhaust gas to drive the turbocharger. Because of this complexity, a full chapter (Chapter 7) is given to scavenging and turbocharging, and you will find a much fuller description there.

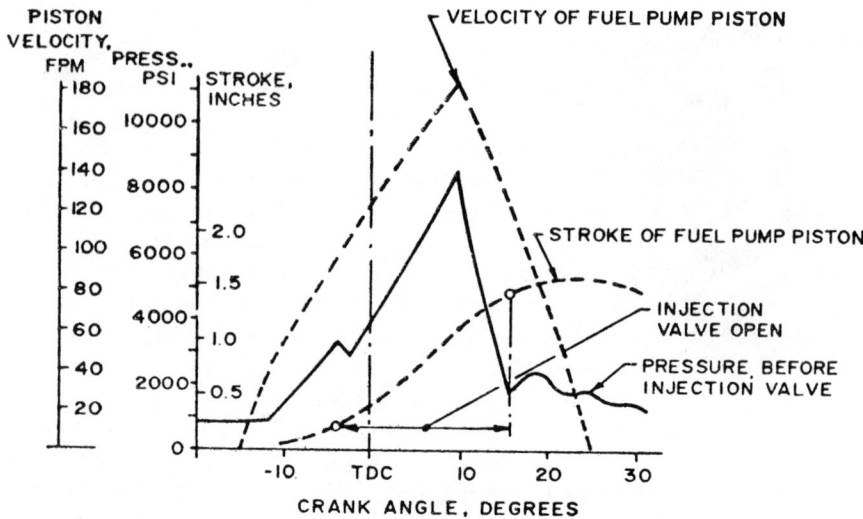

Figure 3.23 Map of injection events (from *Marine Engineering;* by permission of Society of Naval Architects and Marine Engineers).

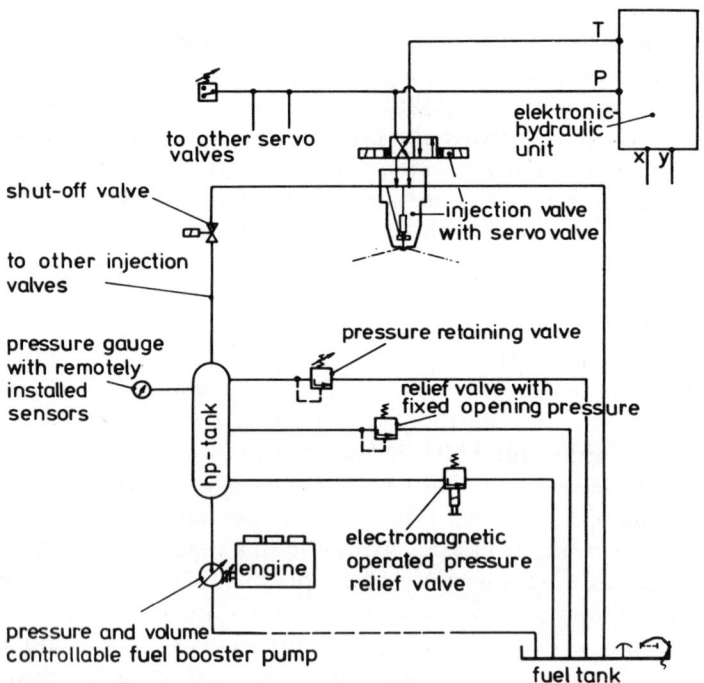

Figure 3.24 Diagram of a fuel system controlled by a Microprocessor (MAN drawing).

3.9 LUBRICATION

Oil is supplied (naturally enough!) to all bearings, usually by an external (that is, not attached to the engine) lube oil pump, with the lower part of the engine bed serving as a sump to which bearing oil drains, and from which the pump takes suction. Some engines, including the one shown in Figure 3.1, also use the lubricating oil as the piston coolant.

The special requirements of the crosshead bearings have been mentioned in Section 3.2. Because high lubricant pressure for them is essential, a lubricator (small pump) driven mechanically by the engine is usually provided for each crosshead.

Low speed engines are unique among the diesel family in that they lubricate their cylinder walls with oil designed for this particular service, and different in its properties from the bearing oil. The ports through which the cylinder oil enters have been noted in Figure 3.9. These ports are supplied by lubricators, one per cylinder, driven mechanically by the engine. Several of these, and their common drive shaft, are visible in Figure 3.1.

Lubricant properties are discussed in Chapter 8, and the lubricating system external to the engine is treated in Chapter 10.

3.10 COOLING

A large fraction of the energy released by fuel combustion inevitably enters the engine structure as heat that must be continuously removed by circulating coolant. The cooling objectives are to keep metal temperatures below the level at which strength is impaired, to reduce temperature gradients in order to reduce thermal stresses, and to keep cylinder surface temperatures within a range that avoids both hot deposits and cold corrosion (points to be amplified in Chapter 8). Lubricating oil picks up heat in the bearings and so must be cooled. Lube oil is also used as the piston coolant in some engines. In addition, the combustion air is cooled before it enters the cylinder, this to increase its density.

Coolant passages or connections in cylinder liners, cylinder covers, pistons, and injectors have been noted in previous sections, and illustrated in Figures 3.4, 3.5, 3.8, 3.9, 3.10, and 3.22. The sliding pipes or swing joints by which coolant is led to and from the moving pistons are especially worthy of notice (Figures 3.4, 3.5, and 3.8). In Figure 3.1 the piston coolant path (oil is this case) is also partially visible.

The combustion air is usually cooled in an air-to-water exchanger, with seawater being the coolant. Water which circulates within the engine (through cylinder liners, covers, etc) is always fresh water in a closed circuit, with exchange of heat to seawater in external heat exchangers. The last word is plural, since several circuits are likely. For example, the water circulating through the fuel injectors faces the hazard of possible oil leakage, and so is isolated from the other coolant services. The lube oil, whether used for piston cooling or not, must also be cooled by external heat exchangers. All of the "externals" mentioned here are being regarded as components of the engine auxiliary systems and hence their discussion is reserved for the auxiliary chapter (Chapter 10).

3.11 CLASSIFICATION SOCIETY RULES ON ENGINE CONSTRUCTION

The several classification societies (American Bureau of Shipping, Lloyd's Register, etc) publish rules governing marine diesel engines. Here are noted in condensed and paraphrased form the requirements of one of the societies (American Bureau) [American Bureau of Shipping (1978)] as they apply to the topics of this chapter.

1. Crankcases (the sump space in the bedplate) are to be ventilated.
2. Explosion relief valves are required in the crankcases of all engines of bore greater than 200 mm (this includes all low speed engines), with size and location of these valves being specified. These valves generally have the appearance of small access ports, spring loaded, in the sides of the engine frame. None are visible in Figure 3.1.
3. Crosshead engines (again, all low speed engines are included in this) with scavenge air spaces openly connected to the cylinder (via ports, that is) must have a permanently connected fire extinguishing system in those spaces.
4. Governors must be installed to prevent speed in excess of 115 percent of rated speed.
5. General integrity of the bedplate is specified ("rigid construction, oiltight, and provided with sufficient bolts to secure to ship's structure").
6. Cylinder relief valves, set to relieve at not more than 40 percent excess over firing pressure, are required on reversible engines, and on engines using air for starting. Either of these includes all low speed engines.
7. Minimum diameters of crankpins and crankshaft journals are specified as functions of cylinder diameter, maximum firing pressure, space between bearings, power, engine speed, and material properties.

8. Minimum crankshaft web proportions and dimensions are specified as functions of crankpin diameter.

3.12 REFERENCES

American Bureau of Shipping (1978), *Rules for Building and Classing Steel Vessels.*
Illies, K (1971), "Low Speed Direct Coupled Diesel Engines," Chapter VIII of *Marine Engineering,* Society of Naval Architects and Marine Engineers.
THE MOTOR SHIP, London, monthly.

3.13 NOTATION FOR CHAPTER 3

B&W	Burmeister & Wain
FPM	feet per minute
GMT	Grandi Motori Trieste
hp	horsepower
MAN	Maschinenfabrik Augsburg Nurnburg
mm	millimeters
MPa	Megapascals
rpm	revolutions per minute
PSI	pounds per square inch
t	tonnes
TDC	top dead center

Chapter Four

ENGINE OPERATIONAL CHARACTERISTICS

Section 1.3 has noted that integration of an engine into a propulsion plant requires a greater engineering effort than might be suspected from the "package" nature of the engine, this because of the many interactions of the engine with the balance of the plant. Understanding and accommodating these interactions begin with knowledge of the engine characteristics. For an elementary example, the design of the ship's fuel tankage obviously depends on how much fuel is to be consumed, and a fundamental ingredient in that quantity is the specific fuel consumption of the engine or engines.

The specific fuel consumption is thus a "characteristic" in the sense being used here. Generally, then, this word implies the value of any engine parameter that can be of interest. In many instances, however, these values are not constants, since they are functions of other parameters that are manipulated in operation, or that change because of environmental change, because of wear, and the like. For example, any parameter that is a function of engine speed changes as the speed of the ship is varied (for the fixed-pitch propeller case, at least). Therefore, one often finds that the characteristic is not a single number, but is a curve expressing the relationship of one parameter to another.

FUEL CONSUMPTION CHARACTERISTICS 55

4.1 TORQUE AND POWER CHARACTERISTICS

Chapter 2 has shown that torque is proportional to mean effective pressure, that the value of the latter is a function only of the amount of fuel injected during each piston cycle, and that this amount is not a function of speed. It follows that torque is not a function of engine speed; the torque-speed characteristic is therefore a horizontal straight line. Since power is the product of torque and speed, the corresponding power characteristic is a straight line of slope proportional to the torque and to the brake mean effective pressure. This exposition is a simplified one, for it does not account for possible change with speed of friction torque, nor for a possible variation in the ability of the turbocharger to supply the combustion air. Nonetheless, the torque and power characteristics for practical use are taken to be these straight lines. Most frequently seen are the power-rpm curves, these being the radial straight lines with slope jointly proportional to torque and bmep. They appear this way throughout the engine literature, including the subsequent discussions in this book (particularly Chapter 5). A significant point is that the power required by the marine load—the screw propeller—falls so rapidly as rpm declines that there is a large excess of engine torque available at low rpm, making the exact magnitude of this torque of small consequence in the usual design environment. See Figure 5.4 for illustration.

4.2 FUEL CONSUMPTION CHARACTERISTICS

The specific fuel consumption (or "fuel rate") of an engine is a vital concern, for indeed, the superiority of the diesel in this parameter over all of its competitors is a major factor in its widespread application. The value is commonly published by engine builders as a single number, which is nearly always close to a value of 200 gr/kWh for all low speed engines.

Now, it is an attractive feature of diesel engines that their specific fuel consumption does not increase greatly as load is decreased from rated conditions. On the other hand, the specific consumption is not constant, and its change over the possible range of speeds or loads may be significant; hence fuel consumption data is sometimes published in the form of a contour map showing consumption at all combinations of speed and power. Figure 4.1 is an example.

Marine engine builders frequently show the fuel characteristic as a single curve, specific fuel consumption as a function of either rpm or power or mep along a fixed-pitch propeller characteristic. This is expedient and satisfactory for the usual application because the engine will

56 ENGINE OPERATIONAL CHARACTERISTICS

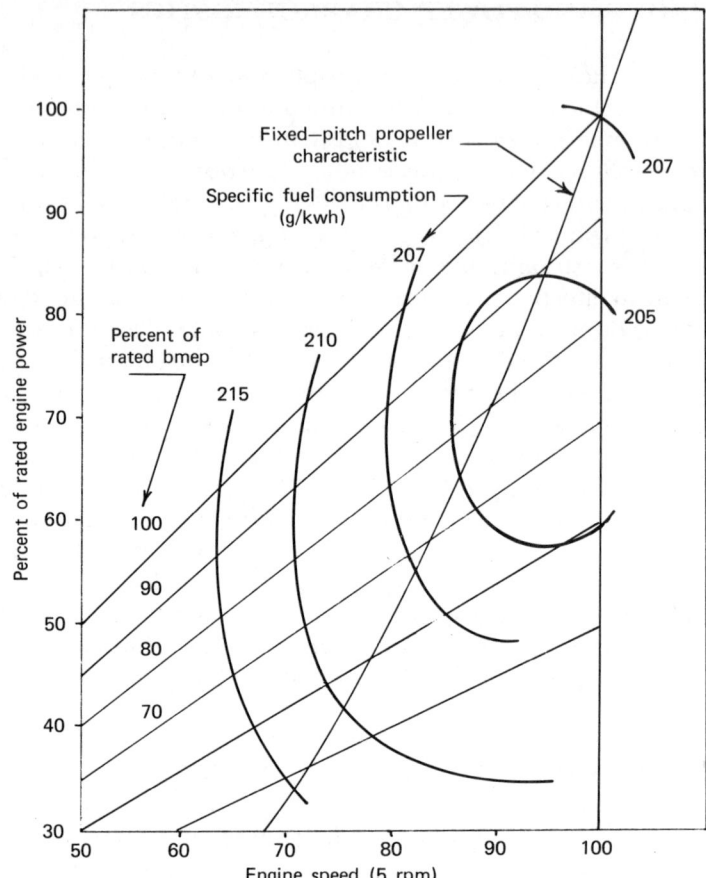

Figure 4.1 Typical fuel rate map, shown on an engine power–speed plane.

operate only along this line in service. A propeller characteristic is included in Figure 4.1; the curve referred to in this paragraph can be constructed by replotting the intersections of the propeller curve with fuel rate contours. Figure 4.2 includes such a single-line characteristic. Although it is for a particular Sulzer model, it is typical of all engines.

However the fuel consumption data is presented, it nearly always originates in shore tests of one engine of the type represented. A ship designer should make allowances for differences among individual engines and among fuels. In general, these allowances are added to the shore fuel rate to obtain a better estimate of actual ship consumption. The diesel performance bulletin of the Society of Naval Architects and Marine Engineers [Society of Naval Architects and Marine Engineers (1975)] lists

FUEL CONSUMPTION CHARACTERISTICS 57

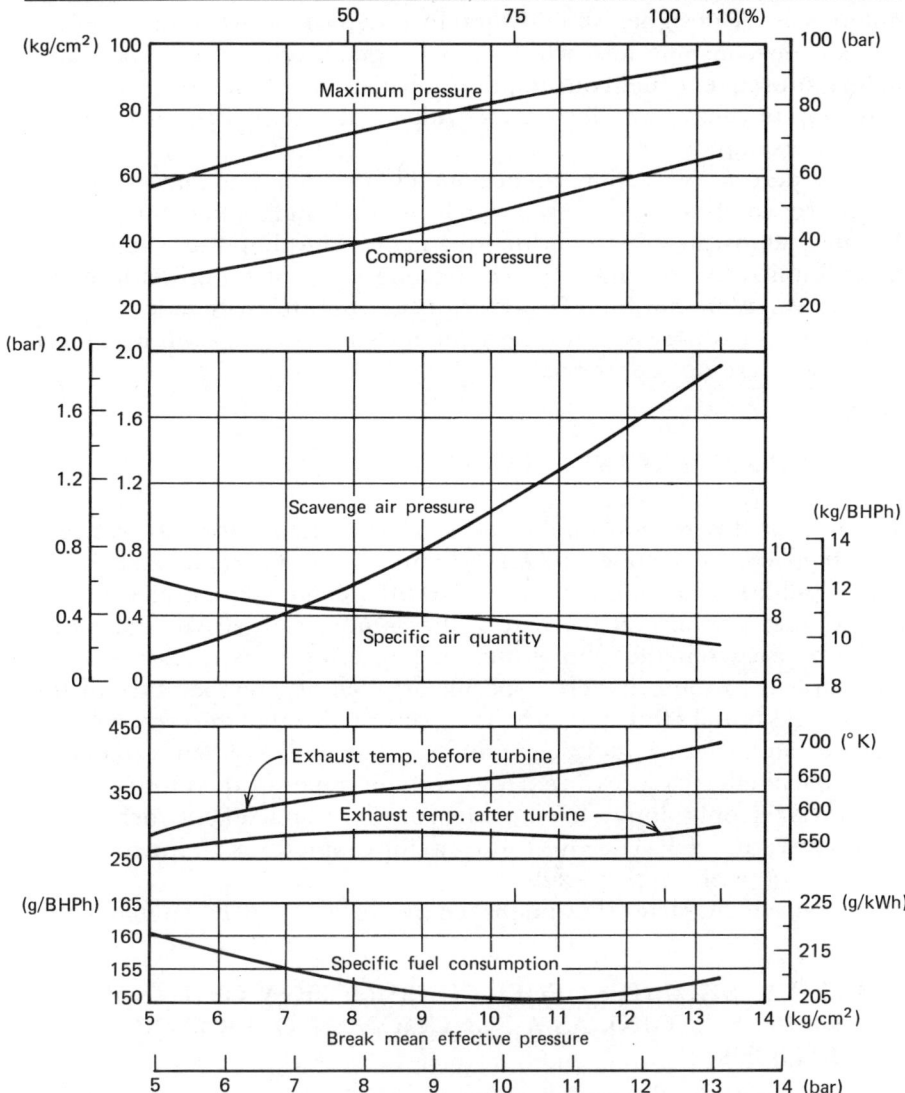

Figure 4.2 Characteristics of a low speed marine engine (Sulzer RND data).

suggested magnitudes of these allowances; the figures quoted in the following paragraphs are from that source.

Differences in fuel rate among engines inevitably occur, and there may be differences caused by lesser expertise in the ship than in the builder's shop where the published rate was measured. A "guaranteed fuel rate" has a margin of 3 to 5 percent added to the test bed rate to allow for these

58 ENGINE OPERATIONAL CHARACTERISTICS

differences. A designer should therefore determine whether a quoted specific fuel consumption includes this margin. If not, it should be added before subsequent fuel consumption calculations are made.

A shop test may have been made with a distillate fuel ("marine diesel oil," for example), while the ship is to be fueled with "heavy oil." If so, then a 2 percent correction can be added to the shop fuel rate to account for poorer combustion expected with heavy fuel, and a 1 percent addition for the fuel lost in shipboard fuel treatment. The shop rate should also be multiplied by the ratio of lower heating value of shop fuel to lower heating value of ship fuel. However, if the shop test is conducted with heavy fuel, or otherwise with the same fuel that the ship will use, these corrections are not appropriate.

4.3 GENERAL ENGINE CHARACTERISTICS

The title of this section implies those characteristics that are typically published by the engine builders. Figure 4.2 is an example, it being curves published by Sulzer for the RND 90M engine. Quantities plotted are cylinder pressure at the end of the compression stroke, maximum pressure reached in the cylinder, scavenging air pressure (pressure of the air supplied to the cylinders), specific air quantity, temperature of the exhaust gas both before and after its expansion through the turbocharger turbine, and specific fuel consumption. The independent variable is brake mean effective pressure. However, the values of the plotted variables are taken only along a fixed-pitch propeller characteristic such as that shown in Figure 4.1. The effect of rpm, for example, as an independent variable is thereby suppressed.

A survey of characteristics published by other low speed engine build-

TABLE 4.1 SHOP TEST FUEL CONSUMPTION OF THE MAN KSZ 90/160 B ENGINE AS A FUNCTION OF OPERATING CONDITIONS

Ambient temperature, C	45	27	20	27	20
Charge air coolant temperature, C	32	27	20	27	20
Scavenge air temperature, C	50	45	30	45	30
Ambient pressure, Pa $\times 10^{-5}$	1.000	1.000	1.000	1.000	1.013
Fuel lower heating value, kJ/kg	42000	42000	42000	42707	42900
Fuel rate, g/kWh	201	198	196	195	191

Source: [Böhm and Simon (1979)].

ers indicates that turbocharger speed is about the only commonly shown characteristic that Sulzer has omitted from this figure. It is a parameter that typically increases approximately linearly by a factor of about two over the bmep range of Figure 4.2, with a speed of 8000 to 9000 rpm as its maximum value at the highest bmep.

From the viewpoint of the ship designer, some of these characteristics are of immediate value, while others are of only incidental interest. The former category is composed of those that have an effect on machinery components and systems external to the engine itself. The fuel rate characteristic, which is a factor in sizing fuel tankage, and in the operational economy of the ship, is certainly one of these. Otherwise, those characteristics relating to air and exhaust are important to the ship designer because they are fundamental to design of exhaust and supply ducting, and to the design of the exhaust waste heat systems so often used with low speed engines.

Other characteristics are also likely to be important to the designer; those of almost certain interest are the external flows and temperatures of cooling water and lube oil. These are rarely published as curves, but quite commonly as single values at the engine rated power. As an example here, values in Table 4.2 were supplied by Burmeister & Wain for the engine (K90GF) of Figure 3.1.

In looking at this table one might note especially the consumption rate of the cylinder lubricant. The cost of supplying this consumable is not a negligible part of operating costs (for instance, perhaps several 10^5 annually in the 1970s), and so is of great interest in engineering economic studies of propulsion plants. The rate given by B&W and quoted in Table 4.2 is a typical value for low speed engines, although you are likely to see

TABLE 4.2 SEVERAL PARAMETER VALUES FOR THE B&W KG90F ENGINE

Cylinder lube oil consumption	26 kg/cylinder day
System lube oil consumption[a]	68 m^3/hr cylinder
System lube oil flow rate	68 m^3/hr cylinder
System oil inlet/outlet temperature	40 to 50/50 to 60 C
Jacket water in/out temperature	50 to 60/55 to 65 C
Jacket water flow rate	74 m^3/hr cylinder
Jacket water pressure at engine	2.5 kg/cm^2

"System oil" is the oil used for general lubrication (i.e., for all lubrication other than the cylinder walls) and piston cooling. See further discussion in Chapters 8 and 10.
Source: Engine Builder's Data Sheets.

60 ENGINE OPERATIONAL CHARACTERISTICS

other figures in the marine engineering literature. The explanation for the difference is that the rate of this lubricant's consumption can be adjusted by the engine operator. An operator may, for example, use a greater amount of lubricant in the expectation of reduced wear rate for the cylinder liners.

4.4 ENGINE HEAT BALANCE

The preceding section mentions the use of characteristic values in design of waste heat recovery systems that are widely used with low speed diesel propulsion systems. The same information is sometimes presented in the form of a graphic "heat balance," a chart that shows the disposition of the energy provided by the combustion of the fuel. Figure 4.3 is an example, a heat balance chart for the Sulzer RND engine. This scheme of presentation is scarcely of use to the designer since coolant fluid flow rates and temperatures are not given, but is illuminating in that it shows quite clearly the relative magnitudes of energy consumptions and losses.

More extensive remarks on waste heat usage are found in Chapter 10.

4.5 EFFECT OF AMBIENT CONDITIONS ON SOME CHARACTERISTICS

Engine performance and characteristics are affected by ambient conditions. Since air drawn from the atmosphere is the working fluid, the state of the atmosphere must influence processes within the engine. Seawater temperature can also be considered an ambient condition in this context because it interacts with the air in the cooler between turbocharger and engine, and thereby affects the state of the working fluid.

Of the items discussed in this chapter, those relating to the exhaust gas are parameters most noticeably affected. Figure 4.4 gives an idea of the changes in gas flow and temperature as a function of air temperature with brake horsepower constant, and Figure 4.5 gives an idea of the change in gas temperature as a function of air temperature leaving the air cooler, a temperature which is in turn a function of seawater temperature. The phrase "gives an idea" is used because the figures are intended here to represent only general behavior, and not data for a particular engine. The source is Norris [Norris (1964)].

Back pressure, the pressure of the space into which the exhaust gas is discharged, can have similar influence, and Figure 4.6 gives an idea of this phenonenon [source is also Norris (1964)]. Such information is also of

EFFECT OF AMBIENT CONDITIONS ON CHARACTERISTICS 61

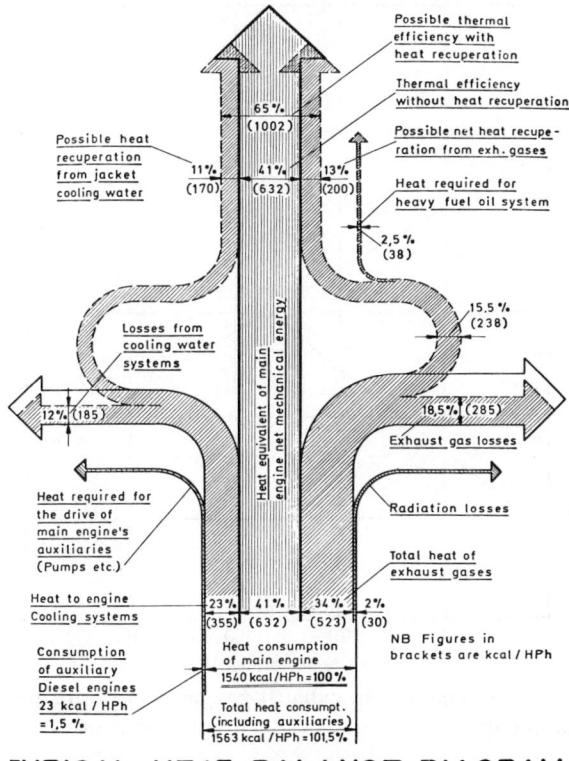

Figure 4.3 Flow of thermal energy ("heat balance") in the Sulzer RND engine (Sulzer drawing).

interest in designing waste heat systems since a waste heat boiler in the exhaust line increases the back pressure.

The Society of Naval Architects and Marine Engineers diesel performance bulletin [Society of Naval Architects and Marine Engineers (1975)] offers a formula for estimating exhaust gas temperature, it being

$$T = T_0 + \frac{(\text{LHV} \times \text{SFC} - 2544) \times M}{\text{SFC} \times C_p \times (R + 1)} \tag{4.1}$$

where T_0 = ambient temperature
LHV = lower heating value of fuel

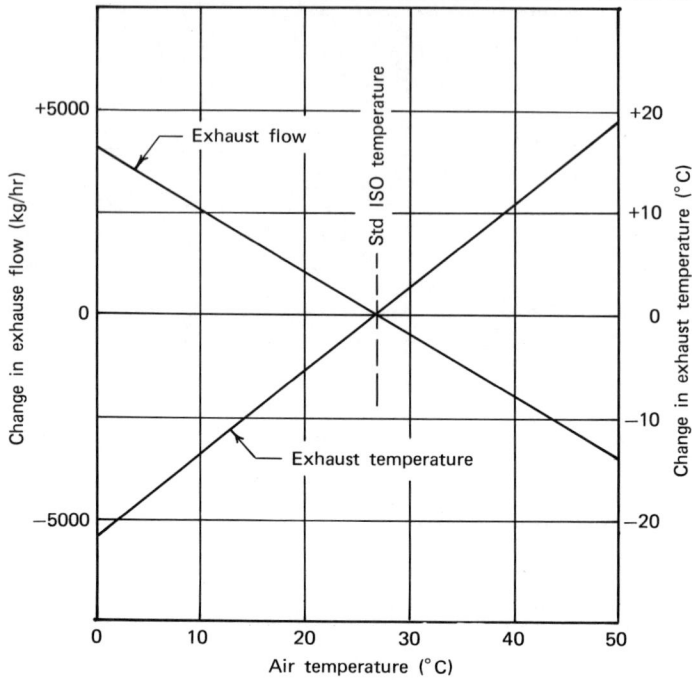

Figure 4.4 Example of changes in exhaust gas mass flow rate and temperature as functions of ambient air temperature [adapted from Norris (1964)].

$\quad$ SFC = specific fuel consumption
$\quad$ C_p = specific heat of exhaust gas
$\quad$ R = air/fuel ratio = 45 at rated power, 55 at half power
$\quad$ M = a constant = 0.55 for distillate fuel, 0.57 for residual
(all units in the English system)

This formula obviously indicates a one-to-one relationship between ambient temperature and exhaust temperature, and gives a relationship of exhaust temperature to load on the engine via the relationship of air/fuel ratio to load.

This formula also indicates the effect of fuel properties on exhaust gas temperature. Using typical values of SFC, LHV, and the two values of M, an example in the source reference shows the residul fuel to produce a 6 F higher exhaust than a distillate diesel fuel.

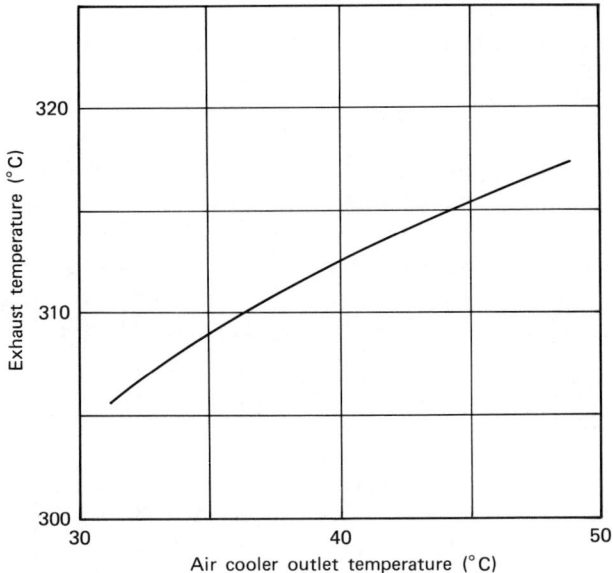

Figure 4.5 Example of change in exhaust gas temperature as a function of air cooler outlet temperature [adapted from Norris (1964)].

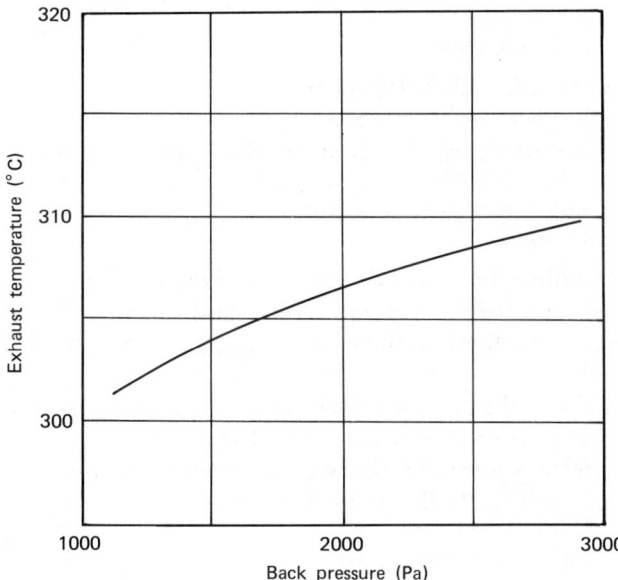

Figure 4.6 Example of change in exhaust gas temperature as a function of exhaust back pressure [adapted from Norris (1964)].

64 ENGINE OPERATIONAL CHARACTERISTICS

4.6 TURBOCHARGER AND ENGINE AIR FLOW CHARACTERISTICS

The engine has a head-flow characteristic representing its action as the load for the turbocharger compressor. In selecting a turbocharger for an engine, the engine designer must match this characteristic to the corresponding compressor characteristic. Although this is not the direct concern of the ship designer, some knowledge of it aids in understanding the behavior of the engine, and a discussion is therefore included in the turbocharger chapter, Chapter 6. See, in particular, Figure 6.10.

4.7 REFERENCES

Norris, Alan (1964), "Developments in Waste Heat Systems for Motor Tankers," *Transactions,* Institute of Marine Engineers, Vol 76, pages 397-429.

Society of Naval Architects and Marine Engineers (1975), *Marine Diesel Power Plant Performance Practices*, Technical and Research Bulletin 3-27.

4.8 NOTATION FOR CHAPTER 4

bar	bar (unit of pressure)
C	degrees Celsius
C_p	constant pressure specific heat
g/BHPh	grams per brake horsepower hour
g/kWh	grams per kilowatt hour
kg	kilogram
kJ/kg	kilojoules per kilogram
kPa	kilopascal
kg/cm^2	kilograms per square centimeter
LHV	lower heating value
M	a constant
mep	mean effective pressure
m^3/hr	cubic meters per hour
R	air/fuel ratio
rpm	revolutions per minute
SFC	specific fuel consumption
T	temperature
T_0	ambient temperature

Chapter Five

THE RELATIONSHIP OF ENGINE TO PROPELLER

Engine and propeller form a single unit whose characteristics are determined by the interactions of the individual characteristics of the two units. This is especially true when the engine is a low speed diesel, since its direct connection to the propeller shaft precludes independent choices of propeller speed and engine speed. And one should add that the characteristics of the hull affect propulsion characteristics, so that engine, propeller, and hull can also be called a single unit.

A marine designer responsible for engine selection and propulsion plant design must therefore be familiar with propeller characteristics, and with hull influences upon them. Such a person will doubtless rely upon the appropriate texts and courses of study for this knowledge. Here, however, is offered a summary to recall the points essential from the engine standpoint.

The propeller accelerates water sternward (when the vessel is being propelled ahead, of course), thus increasing the momentum of the water. The force equivalent to the momentum increase (Newton's Second Law, remember) is the *thrust*, and the product of thrust and the speed of the water relative to the propeller—its *speed of advance*—is the *thrust power*, as expressed by

$$W_t = TV_a$$

$$\text{THP} = \frac{TV_2}{326} \quad (5.1)$$

for W_t, the thrust power in watts, T, the thrust in newtons, and V_a, the speed of advance in meters/second. Or in English units, THP is thrust horsepower for T in pounds force and V_a in knots.

For the ship, the power is applied in overcoming a resistance R at ship speed V, this power being called *effective* power, or *effective horsepower* in English units. In symbols

$$W_e = RV$$

$$\text{EHP} = \frac{RV}{326} \quad (5.2)$$

The thrust power and effective power are equal, but the individual terms usually are not. Speed of advance is usually less than ship speed, and resistance less than thrust. These are expressed by

$$R = T(1 - t) \quad (5.3)$$

$$V_a = V(1 - w) \quad (5.4)$$

where t is *thrust deduction*, and w the *wake fraction*. The ratio $(1 - t)/(1 - w)$ is the *hull efficiency*, symbol η_h, and thus by equations (5.1), (5.2), (5.3), and (5.4),

$$W_e = W_t \eta_h$$

$$\text{EHP} = \text{THP} | \eta_h \quad (5.5)$$

The thrust power, that power which is applied to the water by the propeller, is less than the *delivered power*, symbol W_d or DHP, delivered by the shaft to the propeller. The ratio is *propeller efficiency*, symbol n_p, and this is the product of two efficiencies η_o and η_p. The first of these is the *open water efficiency*, or the efficiency found when the propeller operates without influence of the hull, and the latter is the *relative rotative efficiency*, a correction to no incorporating the hull influences. The delivered power is less than the *shaft power*, symbol W_s or SHP, by the amount of power loss in the stern tube bearing and seals; the ratio of delivered power to shaft power is the *mechanical efficiency* η_m.

The shaft power, that which the engine produces at the inboard shaft seal, is related to effective power, that which calculation or model test indicates is required to drive the ship, by equation (5.5) modified by the terms introduced in the preceding paragraph. The relationship is

$$W_s = \frac{W_e}{\eta_o \eta_r \eta_h \eta_m}$$

$$\text{SHP} = \frac{\text{EHP}}{\eta_o \eta_r \eta_h \eta_m} \quad (5.6)$$

The power delivered by the engine at its output coupling is in turn higher than shaft power by the amount of frictional losses in the transmission. In the absence of gearing or other speed-reducing devices, the loss is very small, usually negligible. For the low speed engine, its rated output, *brake power* (W_b or BHP) is therefore treated as synonymous with shaft power, and the rated engine output is used as the rated propulsion plant output.

5.1 PROPELLER CHARACTERISTICS

Propeller characteristics can be described graphically in several ways. Most appropriate for the discussion to follow are plots of torque coefficient (K_q) and thrust coefficient (K_t) plotted as functions of the advance coefficient (J). They are defined by

$$K_Q = \frac{Q}{\rho D^5 n^2} \quad (5.7)$$

$$K_T = \frac{T}{\rho D^4 n^2} \quad (5.8)$$

$$J = \frac{V_a}{nD} \quad (5.9)$$

where Q is torque, T is thrust, D is propeller diameter, n is rotational speed, and ρ is water density. Units are chosen so that each of the coefficients is non-dimensional. Open water efficiency (η_o) is usually shown on plots of these parameters. It is related to them by

$$\eta_o = \frac{J K_T}{2\pi K_Q} \quad (5.10)$$

A plot of typical K_t, K_q, and η_o characteristics as functions of J is shown by Figure 5.1. A set of these three curves is given for each of five different

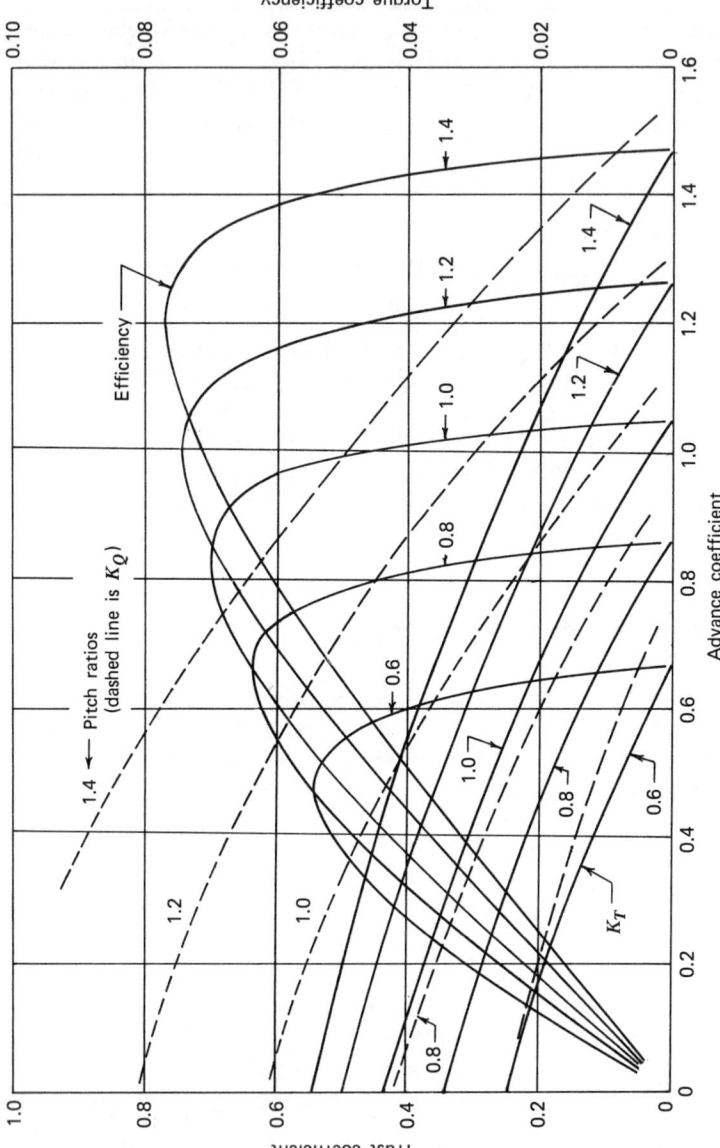

Figure 5.1 Torque coefficient, thrust coefficient, and open water efficiency of a typical propeller as functions of its advance coefficient.

TORQUE AND POWER CHARACTERISTICS OF PROPELLERS 69

pitch ratios over the range 0.6 to 1.4 (pitch ratio = pitch/diameter). These curves represent either a family of fixed-pitch propellers, or a single propeller whose pitch can be varied in service.

5.2 TORQUE AND POWER CHARACTERISTICS OF PROPELLERS

An engine supplies torque to a propeller at some rotational speed. Understanding of combined engine-propeller behavior therefore requires that the propeller characteristics be expressed as a torque-speed relationship, or as a power-speed relationship. A derivation of these from K_t and K_q plots is easily accomplished with the aid of three assumptions: (1) hull resistance is proportional to thrust, (2) resistance (and hence thrust) is proportional to hull speed squared, and (3) hull speed is proportional to propeller rotational speed. The last of these requires that J remain constant as speeds change; if so, then K_q is likewise constant, and torque in consequence must be proportional to n^2. The propeller torque characteristic is therefore a parabola, torque proportional to n^2 also. Since power is the product of torque and rotational speed, the power characteristic is a cubic curve.

Power proportional to cube of speed or torque proportional to square of speed is the principal manifestation of the "propeller law." This law (so-called) is useful for discussions of engine-propeller behavior, but it is not an accurate description of all propulsion situations. For example, its assumptions are highly unrealistic for planing hulls, and are only fair approximations to actuality for displacement hulls driven at high speed-length ratios. Fortunately, the first of these is of no concern here, since the low speed diesel is never found in a planing hull; the second inaccuracy is not sufficient to obviate subsequent discussions and conclusions, and so is ignored here.

The propeller law does not cover transient situations, in which propeller speed and hull speed deviate radically from the assumed proportionality, nor does it cover changes from one steady-state condition to another. As examples of the latter, a change in draft, a change in hull roughness (as from barnacles and the like), or a change in sea state will alter the proportionality between speed squared and resistance. Figure 5.2 illustrates the effect of resistance proportionality changes. Three curves of propeller power vs rpm are shown, for resistances in the ratio 1.0, 1.5, and 2.0 (at any rpm, the resistances have these ratios) for a typical propeller. Each curve is a cubic, and so is obeying the propeller law, but their relative positions are found by manipulation of the K_t, K_q, and J characteristics, and not by any rule of the propeller law.

70 THE RELATIONSHIP OF ENGINE TO PROPELLER

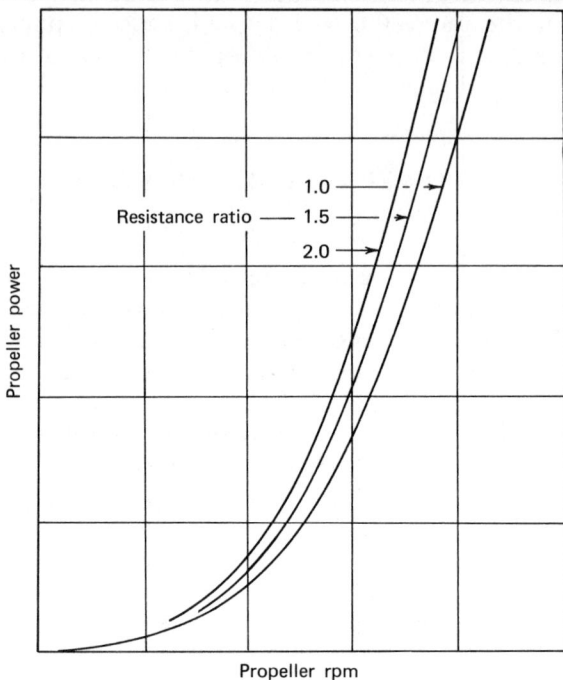

Figure 5.2 Propeller power vs speed (rpm) curves, showing effect of increased resistance.

This law does not account for pitch changes, for such a change also alters the proportionality between speed of hull and speed of propeller. Figure 5.3 illustrates the effect of pitch changes for a typical propeller. Three curves of propeller power vs rpm are shown, for pitch ratios of 1.0, 1.2, and 1.4, with the data being taken from Figure 5.1. Each curve is a cubic, and is individually in accord with the propeller law, but once again, the difference among curves has been found by additional means.

Note the similarity between Figure 5.2 and Figure 5.3. Each is a plot of three cubic power-rpm characteristics, the first figure for different resistances, the other for different pitch ratios. It appears, and truly so, that a change in pitch can offset a change in resistance. For example, when a power-rpm curve shifts to the left because worsened sea state has increased resistance, it can be shifted rightward to its original position by a decrease in pitch (*if* the pitch can be changed, of course). This does not imply that *all* will be the same, for increased resistance will decrease ship speed for a given effective power, and the pitch change may alter propeller efficiency. What it does say is that the characteristic can stay the same,

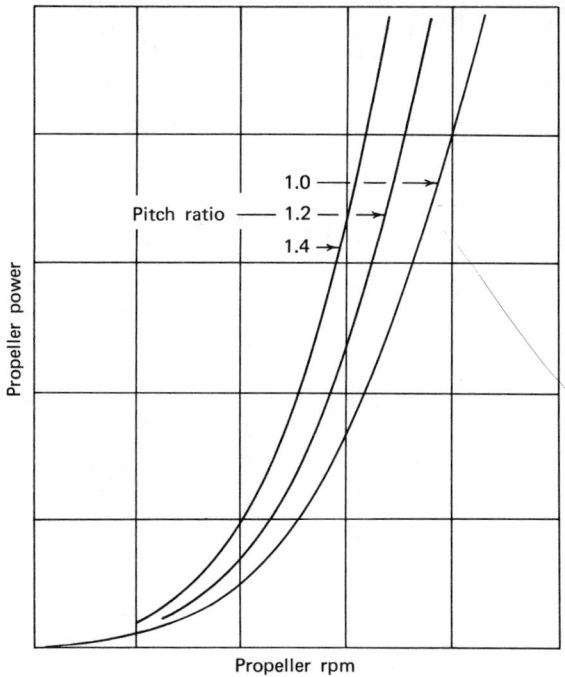

Figure 5.3 Propeller power vs speed (rpm) curves, showing effect of pitch ratio.

that the engine can experience the same relationship between its power or torque and its rpm.

5.3 BASIC MATCH OF ENGINE AND PROPELLER CHARACTERISTICS

Engine output characteristics—the relationships of its torque and its power to its speed—have been noted in Chapters 2 and 4. It is seen there that the ideal characteristics are, respectively, a horizontal straight line and a straight line with slope proportional to torque, and that these ideal lines are close approximations to actual characteristics. They are used in the discussions of this chapter, and likewise in similar discussions found throughout the marine diesel literature.

Figure 5.4 is an elementary plot of power-speed characteristics for a propeller and its driving engine. The operating point is the intersection of the two heavy lines, this being the only place on the plane where power absorbed by the propeller equals that produced by the engine at a com-

72 THE RELATIONSHIP OF ENGINE TO PROPELLER

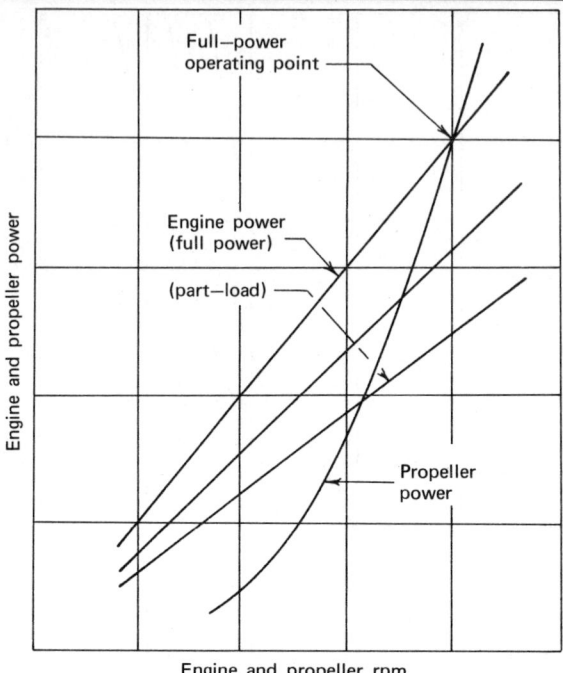

Figure 5.4 The basic concept of engine–propeller matching: operating points determined by intersections of characteristics.

mon rpm. (The small amount of power absorbed by bearings and seals between the two units is neglected here.)

Figure 5.4 also indicated several part-load conditions (light lines). If the ship is to operate at low speed, the fuel injected per cycle is reduced ("fuel rack position" is reduced), resulting in lower mean effective pressure and hence lower torque and lower power at a given rpm. The power-rpm intersections shift accordingly down the propeller curve. By the propeller law, ship speed is changed in the same ratio as the propeller speed. Note that the operating points lie on the propeller curve; engine behavior is thus dictated by the characteristics of its load, the propeller.

The preceding section has shown that the propeller power-speed curves are affected by pitch ratio, with each pitch ratio being represented by a unique curve. Figure 5.5 now demonstrates how this affects the engine. This figure is, essentially, Figure 5.4 modified by the addition of a propeller line for a higher and a lower pitch (diameters the same). If the engine power curve of Figure 5.4 represents the maximum allowed mean effective pressure, and if the rpm is also at its maximum, then these off-pitch propellers prevent the engine from reaching its full power for

BASIC MATCH OF ENGINE AND PROPELLER 73

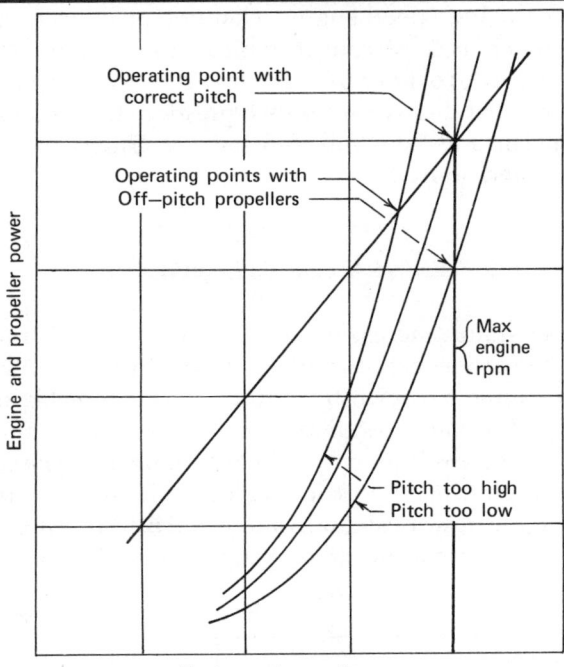

Figure 5.5 The effect on engine power capability of propeller pitch too high and too low.

reasons that Figure 5.5 should make clear. Although this figure simplifies a situation that is actually somewhat more complex, it shows the fundamental fact that the propeller must be selected so that its power or torque characteristic allows the engine to develop the expected output.

The pitch ratio that produces the perfect "match" of engine to propeller may not be the one that gives the highest possible propeller efficiency. This situation is encountered in the design of merchant ships with direct connected engines, most often and most seriously when the ship requires high power at modest speed, typically bulk carriers of 100,000 deadweight tons and above. A typical case might find best propeller design requiring 80 rpm, with low speed engines of suitable power rated at perhaps 110 rpm. The usual solution is to favor the engine, thereby sacrificing some propeller efficiency.

Turbines and medium speed diesels drive the propeller through reduction gears, and since the ship designer has some freedom in selecting the gear reduction ratio, this propeller matching problem can usually be solved by selecting a ratio that allows best rpm for both engine and propeller. The proponents of these engine types cite this freedom as an

74   THE RELATIONSHIP OF ENGINE TO PROPELLER

advantage over the low speed engine, claiming propulsive efficiency differences of as much as 5 percent in typical ship designs. Differences of about half this figure are more usually quoted by low speed engine builders. (For an impartial discussion of this question and its sources, see the analysis by Sinclair and Emerson [Sinclair and Emerson (1968)]; this is a frequently quoted source.)

5.4   DESIGN FOR RESISTANCE CHANGE

The effect of resistance increases on propeller power characteristics has been noted in an earlier section, and illustrated by Figure 5.2. The consequence to the engine is shown by Figure 5.6. Presumably the propeller has been selected to allow the maximum engine output under smooth-water, clean-bottom conditions. But the inevitable increases in resistance occur in service, shifting the propeller curve leftward, with the consequent loss in engine power capability indicated by the figure. Comparison with Figure 5.5 shows that the effect is the same as that of an overpitched

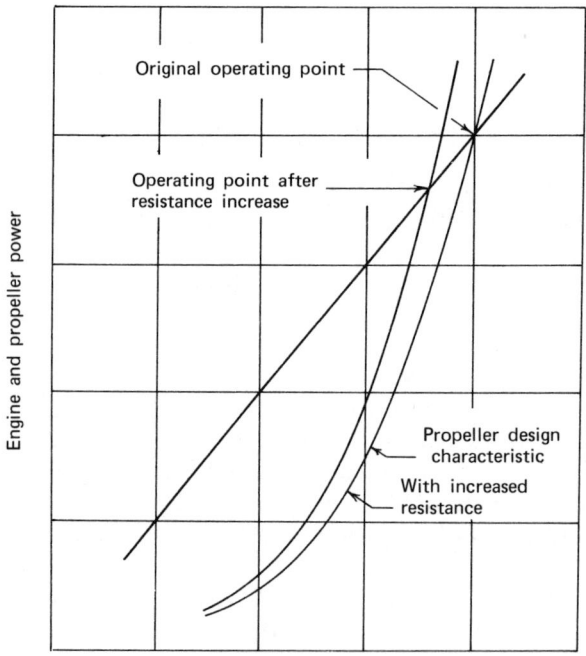

Figure 5.6   The effect on engine power capability of a hull resistance increase.

propeller under smooth water conditions. And this observation suggests a remedy: the propeller should be selected slightly underpitched so that the resistance increase will move the propeller curve *into* the desired position rather than out of it. Although the magnitude of the resistance increase cannot be predicted accurately, and indeed it continuously changes, underpitching is the usual design response to expected resistance increases.

Figure 5.7 illustrates a designer's suggested choice. The point labelled "max continuous" represents the rated power for the engine in question —the power at which the designer intends the engine to run during its normal at-sea service. The crosshatched propeller characteristic is the "trial condition" characteristic augmented by a suitable margin. The smooth-water power ("service hp") is— in this particular case—85 percent of the max continuous. Actual powers will presumably lie somewhere between the service hp and max continuous points, depending upon resistance at the moment. See further discussion of these concepts in Chapter 7.

Figure 5.7 also shows a margin in engine speed; note that "typical max continuous rpm" is designated by a crosshatched line at about 103 percent speed. This allows some increase in rpm above the service hp point when resistance is low.

5.5 CONTROLLABLE PITCH PROPELLERS

The preceding section has spoken of pitch selection and pitch changes in design, and noted that a pitch is to be selected to effect the best compromise of rpm requirements of engine and propeller, modified by concern for resistance changes in service. If the pitch can be changed in service, as it can be with a controllable pitch propeller, the latter problem is bypassed. The propeller may be designed to produce the situation of Figure 5.4, namely the pitch that allows maximum engine output under ideal conditions. When resistance changes occur, tending to shift the propeller curve leftward, the curve can be shifted rightward by pitch decreases to keep it in the same place. The design margin depicted by Figure 5.7 is therefore not necessary. In a typical controllable pitch installation, a pitch governor automatically maintains the propeller curve in its design position by sensing rpm, and reacting to keep this variable constant via an ordered pitch change.

The controllable pitch propeller thus benefits the propulsion plant by allowing it to run at its rated power and rpm in the face of changed external conditions. This does not say, however, that all propulsion conditions are unchanged. For instance, the propeller efficiency changes as

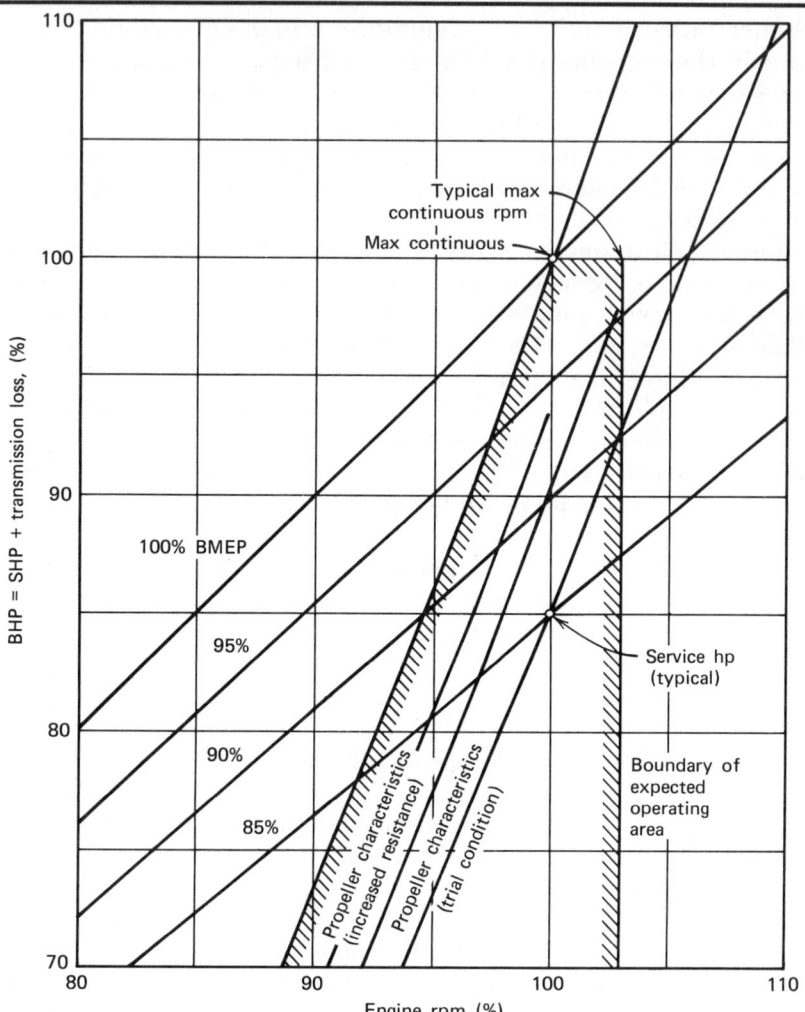

Figure 5.7 Margins in selecting engine–propeller design point [Society of Naval Architects and Marine Engineers (1975), with permission].

pitch ratio changes, and an increase in resistance surely must slow the ship.

A major advantage of the controllable pitch propeller is therefore the capability it bestows upon the engine of running under all conditions at its intended maximum continuous output. Several other advantages related to engine propeller interaction can also be listed.

One is the ability of the propulsion plant to operate at any power-rpm

CONTROLLABLE PITCH PROPELLERS

combination suitable for the engine. (Recall from Section 5.3 that a fixed-pitch propeller allows part-load operation only along the single propeller characteristic.) Changes in power and ship speed can be accomplished at constant rpm, or, at the other extreme, these changes can be accomplished at constant mean effective pressure. The former is of advantage when the engine is driving an auxiliary load whose speed must be kept constant; a constant frequency electrical generator (alternator) is an example. Constant MEP may not be desirable, but values of this parameter higher than those lying along the fixed-pitch path are likely to allow higher engine efficiency. Figure 5.8 illustrates. This figure has contours of fuel rate superimposed on the power-rpm map that occurs frequently in the earlier figures of this chapter. A path from full power to lower powers is laid out through the region of best possible fuel rate. This path can be followed if the corresponding pitch-rpm combinations are chosen. Note, in contrast, that the constant-rpm path to low loads is the worst possible from the standpoint of fuel rate.

Whenever a part-load path is desired, it is usually accomplished via a pitch-MEP-rpm program built into a common engine-propeller control

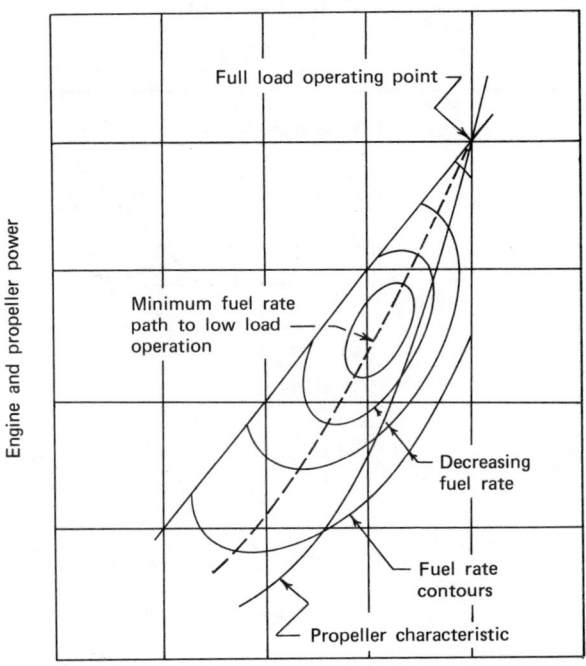

Figure 5.8 A possible best fuel rate path through the lower load region.

78 THE RELATIONSHIP OF ENGINE TO PROPELLER

system. A paper by Schanz [Schanz (1967)] describes such programs and the hardware for their accomplishment; it is perhaps the best description in the English-language literature, and is recommended for further reading.

5.6 NON-PROPULSIVE LOADS

Marine propulsion engines of all types commonly provide power to non-propulsive loads. Generators for ship service electrical power are the most usual. This situation is the least common when the engine is a low speed diesel because it does not have a reduction gear to adjust engine speed to load speed, and generally to provide a ready means of connection to this load. More commonly a low speed engine powers ship service generators through the agency of steam produced from its exhaust heat, a matter not pertinent in this chapter. However, these engines can and do drive generators via speed-increasing gears attached to the propulsion shafting for just this purpose; hence a few words are appropriate here. (See Figure 10.18 for a picture of such a generator-drive gear.)

It is obvious that an auxiliary load subtracts from the engine power

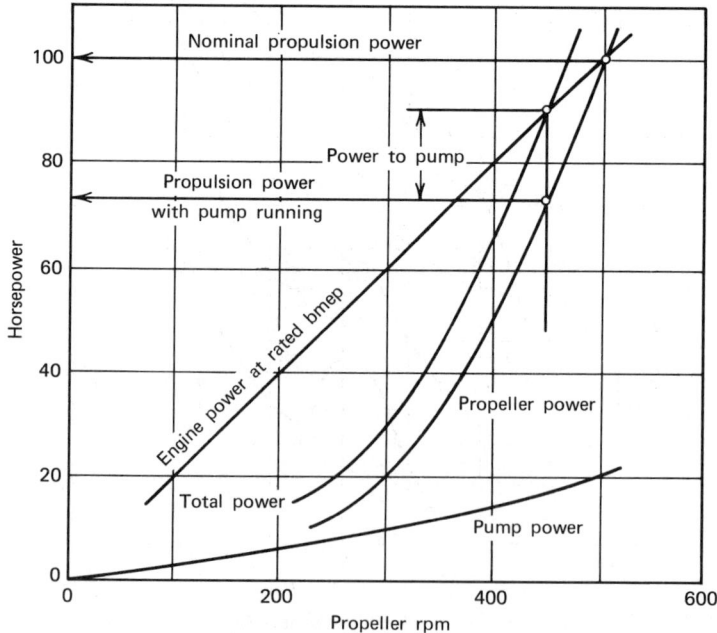

Figure 5.9 Engine and propeller power characteristics with an auxiliary load.

NON-PROPULSIVE LOADS 79

available to the propeller, but additional considerations appear when propeller and load characteristics are examined together. Figure 5.9 illustrates with an example. Here a pump is driven, in addition to the propeller, and the pump power characteristic is plotted on the figure. The original propulsive power is 100. When the pump, additionally, is on the line, the load curve is the sum of propeller power and pump power, and the summed load curve intersects the engine curve at 90, of which 70 is now available to the propeller. If maximum engine output (100, that is) were wanted at the combined load condition, a designer would select a lower propeller pitch to place the curves to the right. Alternatively, a controllable pitch propeller would allow the pitch to be changed when the pump was on the line. The nonpropulsive load therefore has much

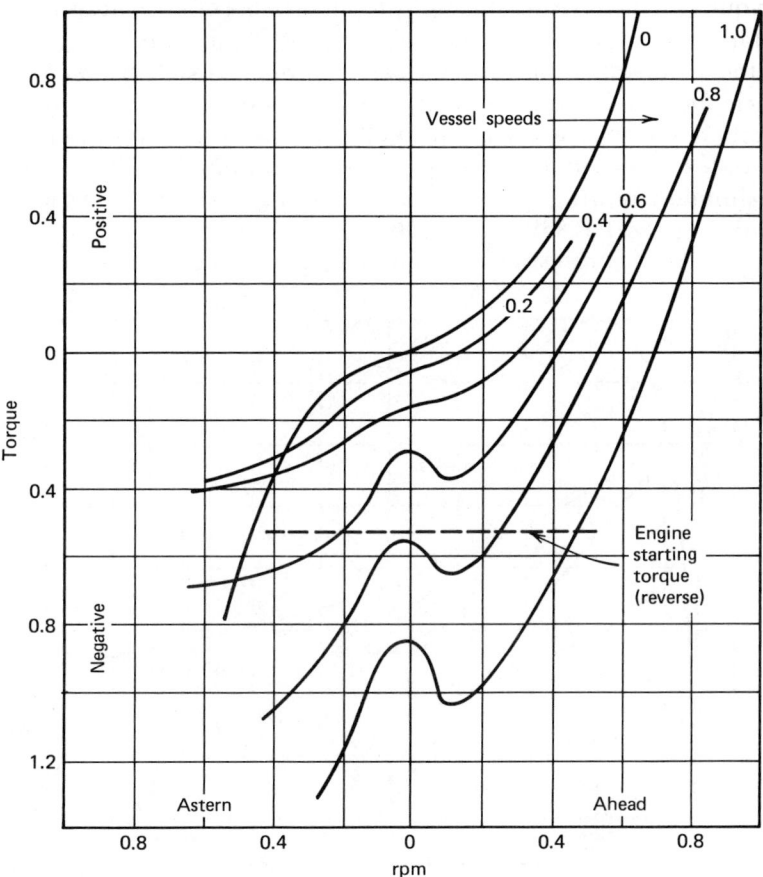

Figure 5.10 Typical propeller torque vs speed (rpm) characteristics during a stopping transient.

80 THE RELATIONSHIP OF ENGINE TO PROPELLER

the same consequences as the resistance increase discussed in Sections 5.3 and 5.4.

5.7 TRANSIENT CONDITIONS

The transient behavior of a propulsion plant also depends on the combined engine-propeller characteristics. A crash stop, for example, calls upon reverse thrust from the propeller, a thrust that depends not only upon propeller characteristics, but upon the ability of the engine to stop and restart the propeller in the reverse direction (speaking here of a direct connected engine with fixed-pitch propeller), and to accelerate it to an appropriate speed.

Propeller characteristics during stopping transients are often shown by a four-quadrant torque-rpm diagram such as Figure 5.10. Note that engine starting torque is also shown, and that when vessel speed ahead has fallen to about 0.7 of its original value (assumed to be 1.00), this torque is sufficient to bring the engine-propeller assembly to rest over the opposition of the "windmilling" torque. If it be assumed for illustration that engine minimum firing speed is 0.2, then firing in reverse can take place when ship speed has fallen to about 0.6.

Figure 5.11 illustrates the contribution of a direct connected reversible

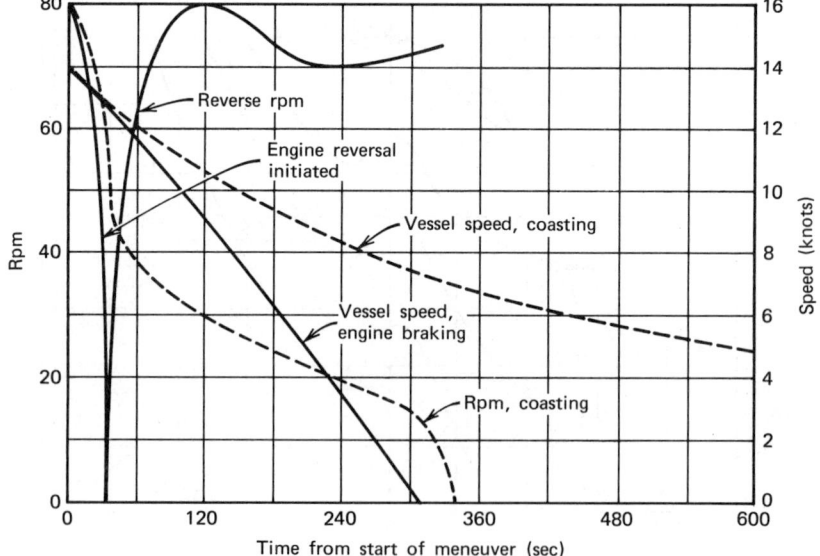

Figure 5.11 Typical stopping time record for a ship with a direct connected diesel engine providing astern torque. Coasting stop shown for comparison.

TABLE 5.1 SOME COMPARATIVE STOPPING TIMES, FIXED-PITCH PROPELLER (FPP) AND CONTROLLABLE PITCH PROPELLER (CPP)

Ship	TDW	Power (BHP)	Starting Speed (knots)	Stopping Time (sec)	Type	Condition[a]
M/S Andorra	12000	12000	19.8	183	CPP	B
M/S Azuma	13150	15000	22.0	160	CPP	B
M/T Esso Fawley	16700	10080	17.0	199	CPP	B
Tanker	18000	8000	15.0	534	FPP	F
M/S Columbialand	24850	11400	15.8	240	CPP	B
M/S Holtefjell	35500	12600	15.5	366	CPP	B
Tanker	33000	11000	15.0	558	FPP	F
Tanker	35000	12500	16.3	582	FPP	F
M/T Sinclair Venezuela	51300	2 × 8400	16.5	286	2 CPP	B
Tanker	47000	15000	16.6	560	FPP	F
Tanker	48500	16000	15.8	630	FPP	F
M/S Nuolja	72500	17600	17.0	420	CPP	B
M/S Nikkala	72500	17600	16.3	426	CPP	F
Tanker	65000	17500	17.0	690	FPP	F
Tanker	79000	22000	15.9	750	FPP	F
SS Fort Henry	12000	6000	20 mph	293	FPP	B
M/V Fort Chambly	12000	6000	20 mph	163	CPP	B
SS Murray Bay	25000	10000	17.75 mph	296	FPP	B
M/V Saguenay	25000	9500	17.75 mph	285	CPP	B

Source: Ridley and Midttun (1970). By permission of *Naval Engineers Journal.*
[a] B = Ballast; F = full load.

engine during the stopping transient of a large tanker. Two cases are shown, one being simply a coasting stop, and the other a stop with engine reversed. The reduced stopping time in the latter case is evident in the figure.

A controllable pitch propeller of the usual design does not require reversal of rotation to produce astern thrust, since its pitch can be reversed. Stopping times are generally shorter than with fixed-pitch because the reverse thrust can be applied continuously in the greatest amount allowed by the engine's ability to produce the requisite torque. Table 5.1 illustrates the superior capability of the controllable pitch propeller in this respect; recorded stopping times for a number of comparable vessels—some with fixed-pitch, some with reversible pitch—are shown. Although exact comparisons are not intended, because of some other differences among the ships, the superiority of the latter propeller is well shown.

5.8 REFERENCES

Ridley, Donald, and Midttun, Ole (1970), "Economic Considerations of the Controllable Pitch Propeller with Diesel Main Propulsion," *Naval Engineers Journal*, Vol 82, 2, pages 71-86.

Schanz, F (1967), "The Controllable Pitch Propeller as an Integral Part of the Ship's Propulsion System," *Transactions*, Society of Naval Architects and Marine Engineers, Vol 75, pages 194-223.

Sinclair, L, and Emerson, A (1968), "The Development of Propellers for High Powered Merchant Vessels," *Transactions*, Institute of Marine Engineers, Vol 80, pages 129-149.

Society of Naval Architects and Marine Engineers (1975), *Marine Diesel Power Plant Performance Practices*, Technical and Research Bulletin 3-27.

5.9 NOTATION FOR CHAPTER 5

BHP, bhp	brake horsepower
D	propeller diameter
DHP, dhp	delivered horsepower
EHP, ehp	effective horsepower
H	propeller pitch
J	propeller advance coefficient
K_q	propeller torque coefficient
K_t	propeller thrust coefficient
mep	mean effective pressure

NOTATION FOR CHAPTER 5

N, n	propeller rotational speed
R	resistance
rpm	rotational speed (revolutions per minute)
SHP, shp	shaft horsepower
T	thrust
t	thrust deduction
V	speed of ship
V_a	speed of advance of propeller
w	wake fraction
W_d	delivered power
W_e	effective power
W_s	shaft power
W_t	thrust power
n_n	hull efficiency
n_o	propeller open water efficiency
n_p	propeller efficiency
n_r	relative rotative efficiency
ρ	density of water

Chapter

Six

SCAVENGING AND TURBOCHARGING

Air in the diesel cylinder is essential both as an ingredient in the combustion process, and as the working fluid. As an absolute minimum for complete combustion, an air-to-fuel ratio of about 15-to-1 is required for the common hydrocarbon fuels. A practical minimum must be somewhat higher because of the impossibility of obtaining a perfect mix of fuel and oxygen during the short time available for this mixing to take place; the lowest ratio that will produce complete combustion in the cylinder is about 20-to-1.

Exigencies of thermal loadings of pistons and other combustion chamber boundaries may require even higher ratios of air to fuel. Generally, the larger the engine, the more air demanded by this requirement. For instance, an automotive diesel at full load may operate close to the approximate 20-to-1 limit just mentioned; a large low speed engine may require about twice this ratio.

Two-stroke engines require a further excess of air for *scavenging*, this being the process of replacing a spent fuel-air mixture with fresh air while the piston is passing near bottom center. Because of the inevitable imperfections in the process, this excess must be provided to compensate for the air which escapes with the spent gas while the exhaust ports or valves are open. Typically, the large low speed engine consumes air at about a 50-to-1 air-to-fuel ratio.

Such, in brief, is the story of the ratios of air to fuel, at least at full power. But further, the magnitude of "full power" depends on the absolute mass of fuel burned, and hence on the absolute mass of air provided to the cylinder. Given that a cylinder represents a fixed volume, the mass of air provided—or forced in—is proportional to its density. Because of the economic merit of obtaining the maximum output from that fixed cylinder volume, a compressor is usually fitted (*always* in the case of the low speed engine) to raise cylinder inlet pressure two to three times atmospheric.

The compressor can take several forms—reciprocating, rotary, turbo—and it can be driven in a number of different ways. Universally, however, the *turbocharger* is used in lieu of all others. (Other compressors are often used to supplement the turbocharger.) This device is a centrifugal compressor driven by a gas turbine whose source of energy is the exhaust gas discharged by the engine cylinders. The characteristics of the turbocharger, its relationship to the engine and to the engine performance, and its several ways of operating in the scavenging and charging process are the major topics of this chapter.

6.1 SCAVENGING PRINCIPLES

As the piston of a two-stroke engine passes near its bottom center position, it uncovers exhaust ports (or in some engines an exhaust valve opens at this time) and inlet ports. The cylinder gas flows out, at first because its pressure exceeds that of the exhaust space, then continuing because it is impelled by the fresh air coming from the scavenging blower or compressor via the inlet ports. By the time the cylinder is again closed to the exhaust and inlet spaces by the upward stroke of the piston, the process of replacing spent gas with a fresh air charge —the scavenging process—should be complete.

Ideally, scavenging totally exhausts the spent cylinder gas, refills the cylinder with a fresh charge that is uncontaminated by residues of the exhaust gas, yet does not waste any of the fresh charge by allowing it to excape through the exhaust ports. Such an ideal would exist only if the incoming air pushed the spent gas toward the exhaust ports without mixing, and if the ports closed just as the last wisp of spent gas passed through. Imperfections intervene, however. Some mixing between the two gases is unavoidable, and indeed is preferable to an extreme situation in which fresh charge short-circuits from inlet to exhaust without disturbing the bulk of the cylinder contents. Perfections of port or valve timing are possible, but unlikely to be attained.

The degree of perfection in replacing the spent gas is expressed by the *scavenging efficiency*, the ratio of fresh air mass actually retained in the cylinder to that ideally retained. The latter quantity is taken to be the mass of a volume of air equal to the cylinder volume at bottom center, with inlet temperature and exhaust pressure. In terms of an air flow rate this efficiency (η_s) is

$$\eta_s = \frac{\dot{m}_a}{NV_d\rho_s[r/(r-1)]} \tag{6.1}$$

where
$\dot{m}_a$ = time rate of retained air flow
N = engine rpm
V_d = displacement volume
ρ_s = density of air at the stated conditions
r_a = compression ratio

The scavenging efficiency is a function of the *scavenging ratio*, this being the ratio of the mass of air supplied to the mass of air at inlet temperature and exhaust pressure that would just fill the cylinder volume at bottom center. This ratio (R_s) in terms of air flow rate is therefore

$$R_s = \frac{\dot{m}_i}{NV_d[r/(r-1)]} \tag{6.2}$$

If scavenging were accomplished ideally, η_s would equal R_s (and $\dot{m}_a = \dot{m}_i$). In practice, η_s is always less than R_s, since some excess of fresh air must be supplied to reach a desired level of scavenging efficiency. For instance, if complete mixing between incoming and exhausting streams occurs,

$$\eta_s = 1 - e^{-R_s} \tag{6.3}$$

(This equation is exact only if cylinder gas and the incoming air have the same temperature and molecular weight, and if the piston remains at bottom center during the process.) Equation (6.3) is plotted in Figure 6.1 for comparison with the ideal case of $\eta_s = R_s$. It is apparent that a large excess of scavenging air must be supplied if a high scavenging efficiency is to be reached. With mixing, $\eta_s = 1.0$ is an unreachable goal, of course, since mixing implies that some exhaust gas remains in the cylinder when the scavenging process is terminated.

The "short circuit" alternative to mixing cannot be so readily put into equation form. Its quantitative consequences can be roughly estimated, however, and shown by a third curve in the same figure. The curve is for an extreme case of severe short-circuiting of air from inlet to exhaust.

The $\eta_s - R_s$ relationship actually found in a particular case depends on several factors. A major one is the path of the scavenging flow as guided by the shape and position of the ports and valves (if any), a factor that is treated in the next section. The effects of piston speed, cylinder size,

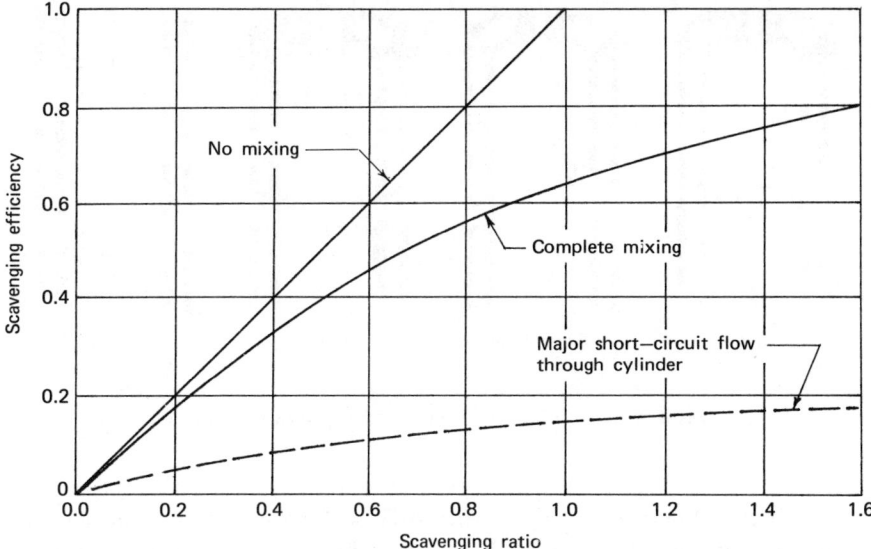

Figure 6.1 Scavenging efficiency as a function of scavenging ratio for several cases [adapted from Schweitzer (1940)].

ratio of port or valve area to piston area, and scavenging pressure can also be identified as having some consequences. The reader should consult texts [Taylor and Taylor (1961)], [Obert (1968)] specializing in internal combustion engine design for amplification of these points.

6.2 SCAVENGING ARRANGEMENTS

High scavenging efficiency is obviously desirable—it maximizes the amount of oxygen in the cylinder, for instance. And low scavenging ratio is likewise desirable, principally because it minimizes the energy devoted to air handling. Pursuit of these goals has led to the development of several arrangements of ports and valves, three of which are used by low speed marine engines; they are *cross* scavenging, *loop* scavenging, and *uniflow* scavenging. These are pictured in Figure 6.2. The implications of the three terms should be clear from the figure.

Figure 6.3 is a diagram similar to Figure 6.1, but with the addition of η_s vs R_s plots for the three scavenging arrangements. Some difference among the arrangements is evident, with uniflow having the highest ratio of η_s to R_s throughout the range shown. This advantage is purchased at the expense of additional mechanical complexity, that of the necessary exhaust valve or of the opposed piston.

88 SCAVENGING AND TURBOCHARGING

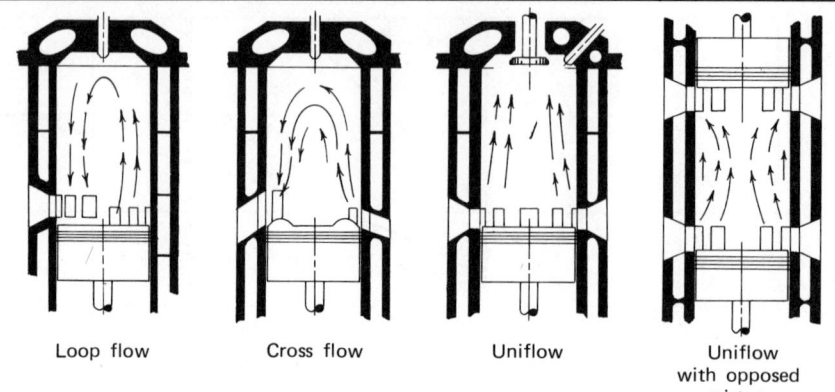

Figure 6.2 The three scavenging arrangements.

Recall that Section 6.1 mentions several factors other than arrangement that can affect the η_s vs R_s relationship. The curves of Figure 6.3 must therefore be regarded only as representative of the three arrangements, and not as presenting the full range of relationship possible.

A major point in a discussion of scavenging is the method of supplying the scavenging air. A pump or blower, driven either by the engine or by a separate power source, is a necessity. When the engine is turbocharged

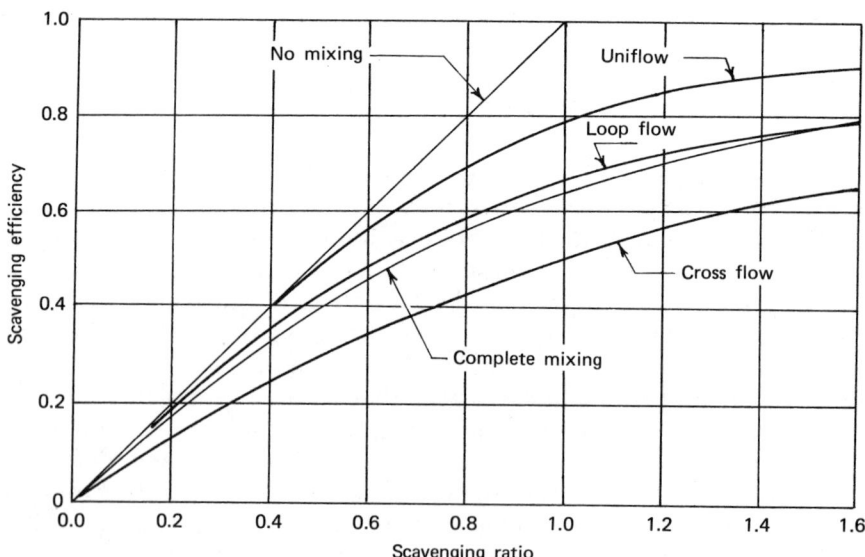

Figure 6.3 Scavenging efficiency as a function of scavenging arrangements [adapted from Schweitzer (1949)].

GENERAL PRINCIPLES OF TURBOCHARGING 89

— as is the case with all engines of interest here—the turbocharger is the principal agent of air supply, even though it is often supplemented by other devices. The discussion of air supply is consequently to be deferred for the moment, to be included in the treatment of turbocharging that begins in the next section.

6.3 GENERAL PRINCIPLES OF TURBOCHARGING

The higher the pressure of air supplied for scavenging, the greater will be the density of air trapped within the cylinder when all ports or valves have closed. And the greater this density, the higher the mean effective pressure of the cylinder. In pursuit of high output, all low speed marine engines (all constructed since the mid 1960s, at least)are provided with scavenging air at pressures much greater than needed for scavenging alone. For example, engines of the 1970s typically have mean effective pressures in the neighborhood of 1200 kPa (12 bar, or 174 lb/in^2),with consequent required compressor compression ratios of about 3-to-1 (scavenging pressure three times atmospheric).

Numerous ways of providing the compression are possible since several types of compressor (reciprocating, rotary, turbo) and several sources of energy are usable. Among the practicable alternatives are reciprocating compressors driven by linkages from the crossheads, rotary or turbo compressors driven by motors energized from the ship service electrical system, the undersides of pistons acting as reciprocating compressors, and turbocompressors driven by exhaust gas turbines. All of these are used, but the last—the *turbocharger*—is universally applied; the others are used only to supplement the turbocharger.

The turbocharger offers the benefit of using energy present in the engine cylinder which is unavailable to the piston, and doing so in a device that is compact and mechanically simple. Figure 6.4 introduces the concept in an elementary way: a centrifugal compressor is driven by an axial flow turbine. The air, compressed, fills a chamber that represents an engine cylinder at a density determined by local pressure and temperature; it exhausts via a turbine nozzle thus driving the turbine. The peak pressure in the cylinder is, of course, many times that provided by the compressor, but the higher pressures occur only while the cylinder is isolated by closure of its ports or valves. Since the turbocharger supplies several cylinders that open to it in turn, a path from compressor to turbine is available almost continuously through a chamber in which the pressure is effectively that of the compressor outlet, less flow losses.

Figure 6.5 (taken from an engine builder's advertisement) pictures the compressor, turbine, and engine cylinders in a somewhat more realistic configuration than Figure 6.4. A feature that is strongly evident here, and

90 SCAVENGING AND TURBOCHARGING

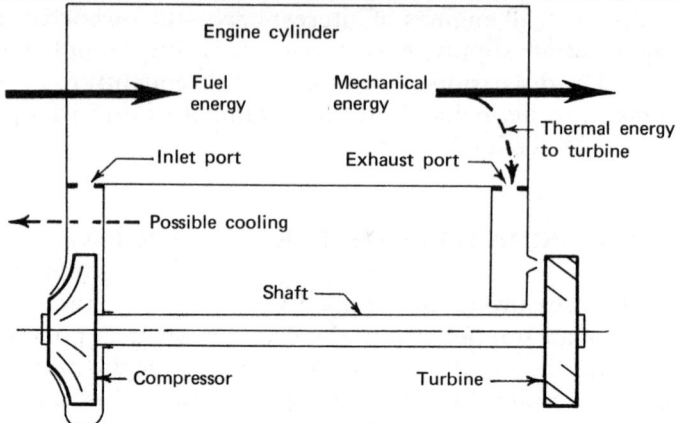

Figure 6.4 General concept of turbocharging.

also indicated in Figure 6.4, is the cooling of the compressed air that occurs before the air reaches the inlet ports. This is a universal practice that carries the benefit of an increased density at the expense of small flow losses in the cooler.

If the compression, flow from compressor to turbine, and expansion back to atmospheric pressure were all reversible processes, the provision of scavenging air at high pressure would be obtained at no net energy input. Put simply, the compressor would supply the compressed air to run the turbine, while the turbine would supply the torque to turn the compressor. Along the air flow path between the two, the engine cylinder would 'borrow' the air for its use. Indeed, the turbocharger would have an excess of mechanical energy since the air in passing through the cylinder must accept the thermal energy not available to the piston (the

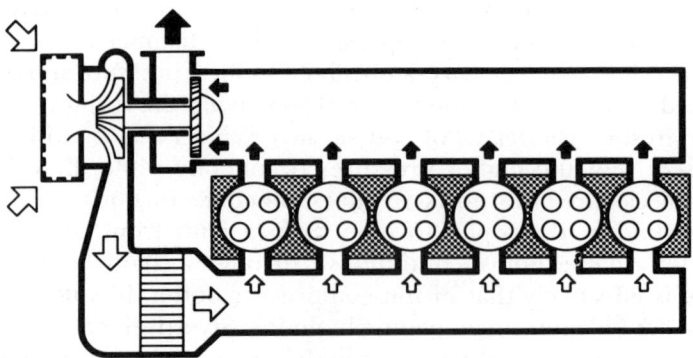

Figure 6.5 Further illustration of turbocharging (MAN drawing).

second law of thermodynamics dictates that the diesel working fluid be hotter at the end of the expansion stroke than at the beginning of the compression stroke). The ideal is far from being attainable, for in reality, both compressor and turbine are less than 100 percent efficient, and throttling losses occur in ports and valves as the air flows from compressor to turbine via the cylinder. Nonetheless, the deficiencies that these represent can be made up by that thermal energy unavailable to the piston.

A conception of the engine-turbocharger interaction is presented on the enthalpy-entropy plane of Figure 6.6, essentially a Brayton cycle with irreversibilities. The compression and expansion lines are shown with irreversibilities evident; a constant enthalpy line (B to C) indicates throttling losses, and a constant pressure line (C to D) describes the addition of heat from the diesel process. For self-sustained operation, the turbine work (D to E) must equal the compressor work (A to B). It should be evident from the figure that the less efficient are the turbine and compressor, the greater must be the energy input from the cylinder. Nonetheless, as long as the temperature level of point D need not be above that existing in the cylinder when the exhaust ports open for scavenging, the

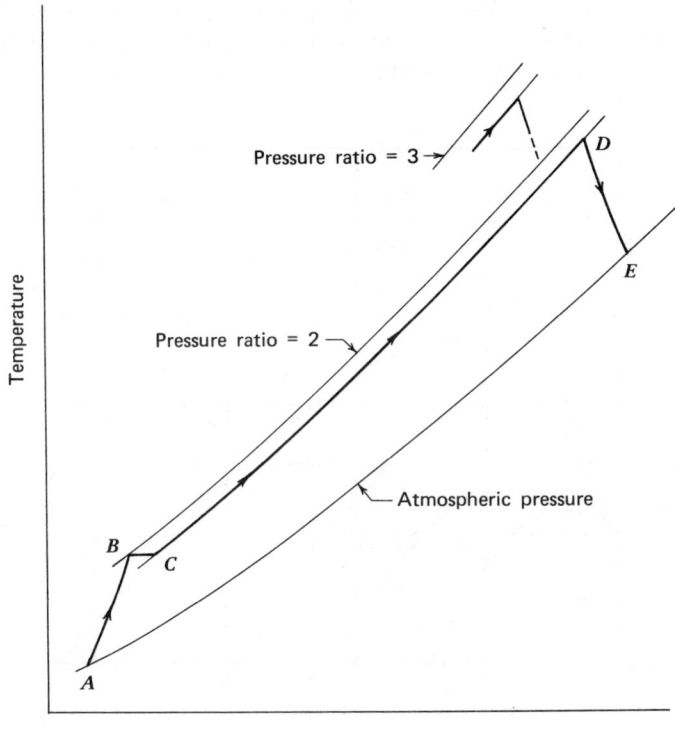

Figure 6.6 Turbocharger processes on a temperature–entropy plane.

energy to run the scavenging-plus-sudevice is essentially free, that is, does not exceed the amount unavailable to the piston. Additionally, some kinetic energy from the cylinder is used by the turbine when the turbocharger system is designed for the so-called "pulse" operation. See further discussion in Section 6.5.

Figure 6.6 is drawn for a pressure ratio of 2-to-1. In addition, it includes a fragment of a cycle characterized by a 3-to-1 pressure ratio, but otherwise with the same component efficiencies and the same fractional throttling loss. The significant point for the higher pressure ratio is that the peak temperature is higher than before. Although a calculation will show that the thermal energy input is not greater, it occurs at a higher temperature level, and thus may demand energy that *is* available to the diesel cycle (that is, require that the cylinder gas exhaust to the turbine before it has done all possible work on the piston).

One concludes from Figure 6.6 and the accompanying discussion that the efficiencies of compressor and turbine are the key to the successful application of the turbocharger to the diesel cylinder: the higher these efficiencies, the higher can be the scavenging pressure (hence density of air in the cylinder, hence engine mean effective pressure likewise) without using energy that should be applied to the piston. Figure 6.7 gives an indication of what the efficiencies should be for a reasonable cylinder exhaust temperature in typical engines. It shows the ratio of compressor

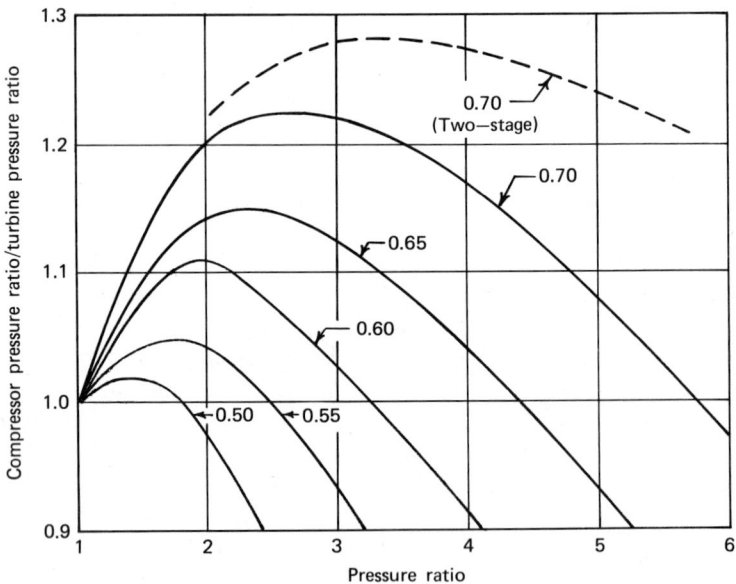

Figure 6.7 Required turbocharger efficiencies as functions of pressure ratios [adapted from *Brown Boveri Review* (1977)].

pressure ratio to turbine pressure ratio as a function of the former, with several levels of combined efficiency (product of compressor efficiency and turbine efficiency). Engines of the late 1970s require a compressor pressure ratio of about 3-to-1, with consequent compressor/turbine ratio in the range 1.1 to 1.2. The figure indicates that a combined efficiency in the neighborhood of 0.65 is required [*Brown Boveri Review* (1977)].

It should be apparent that a great deal more energy from the engine cylinder *can* be assigned to running the turbocharger than a designer of the optimal diesel engine might wish. The assignment is made by opening the exhaust ports or valves earlier in the power stroke of the piston. At the extreme they would be opened as soon as enough energy had been absorbed by the piston to carry out the compression stroke, leaving none to turn the output shaft; the turbocharger then would have sufficient excess energy to turn an output shaft. If it were reasonably efficient, it might well turn the propeller. The engine cylinder in such an arrangement would be only a supplier of gas in a gas turbine propulsion plant. Such is the essence of the 'free piston' gas turbine. Students of marine technical history may recall that this mating of diesel and gas turbine elements was tried in several ships during the 1950s, but was found to be inferior to both conventional gas turbine and conventional diesel propulsion.

It remains, therefore, that only very small sacrifices in diesel cylinder output are made to turbocharger requirements. When these requirements are greater than the purely "waste" energy of the cylinder, it is usually found to be more economical to provide them from the mechanical output of the engine, that is, by some form of mechanically driven compressor to supplement the turbocompressor. Indeed, such devices are typical of the low speed engines. Some techniques of supplementing the turbocharger are discussed in Section 6.5.

The preceding paragraphs explain that the "free" energy from the cylinder may be insufficient to power the turbocharger, and imply that the rated (full power) condition is being discussed. The discussion must be extended to low-load conditions, for these present quite a different situation with respect to turbocharger energy supply. Reflect that the air consumption per cylinder cycle is approximately the same at all loads; the air-fuel ratio rises as load goes down because it is the fuel, not the air, that is reduced. Consequently the temperature of the cylinder gas at any point in the piston stroke—a point such as the opening of the exhaust— declines with load. Although the load on the turbocharger also declines (note that the flow per unit *time* declines as engine speed goes down, even though the flow per cycle stays the same), the availability of energy in the cylinder gas is less (point D in Figure 6.6 is lower). Even if the turbocharger requires no mechanical assistance at full power, it is likely to need it at low powers. Several assisting techniques are discussed in Section 6.6.

6.4 TURBOCHARGER CHARACTERISTICS

Figure 6.8 gives two illustrations of turbochargers. The top one is an exploded view (casings pulled apart to expose the rotor assembly) of a Napier unit, with compressor and turbine casings partly cut away. The other is a cross-sectional drawing of a Brown Boveri unit. In both cases, the compressor and its air inlet lie to the left.

Figure 6.9 presents typical compressor characteristics, these being a plot of pressure ratios vs corrected mass flow rates, with rotational speed as a parameter. Contours of compressor efficiency are also included. Also

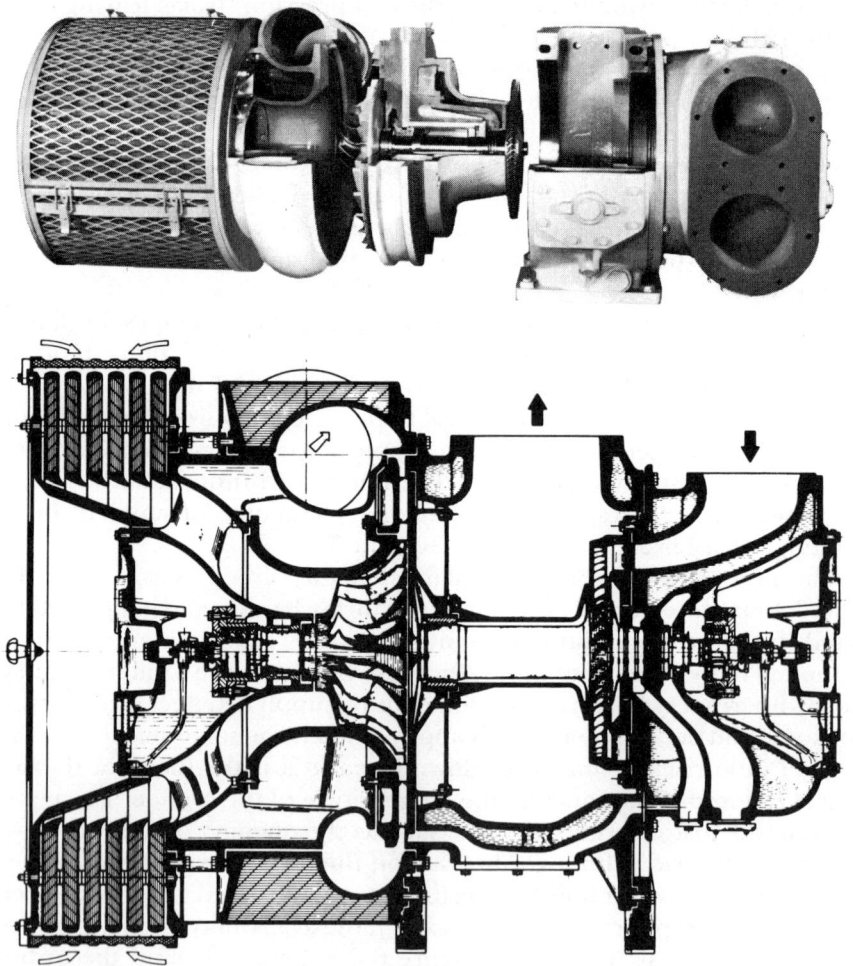

Figure 6.8 Turbocharger illustrations. Top: Napier unit, exploded and cutaway view. Bottom: Cross section of Brown Boveri unit.

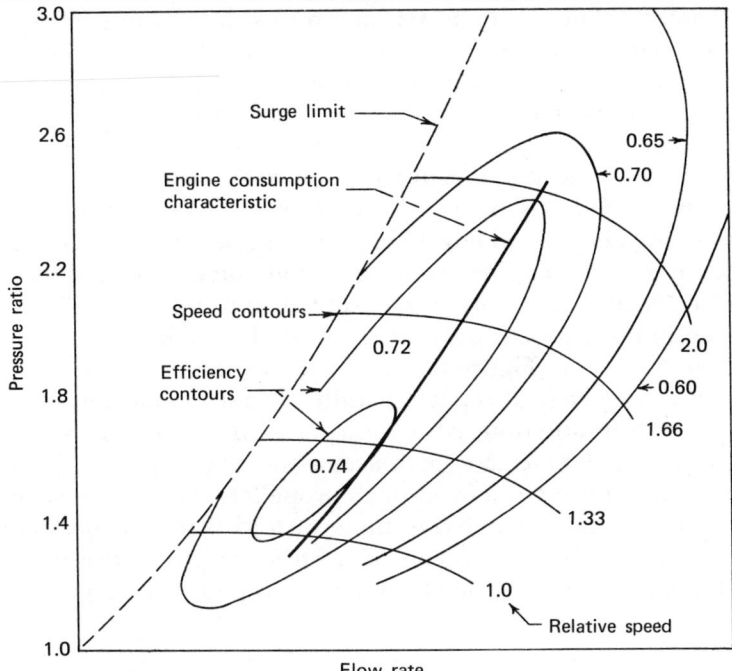

Figure 6.9 Typical compressor characteristics.

shown is a typical air consumption curve for an engine at a constant rpm (or for the group of cylinders served by the turbocharger). Since both compressor curves and engine curve are intended to be typical only, not representing actual machines in either case, the abscissa scale is not numbered. Flow magnitudes depend on the size of compressor, and on the size and configuration of the flow path through the cylinders. The position of the engine curve on the plot is therefore neither fortuitous, nor does it lie in its seemingly favorable location automatically. Rather, its position is a result of careful matching of turbocharger to engine by the engine designer. In essence, this means that a compressor size is chosen to meet the air flow characteristic of the engine or of the group of cylinders served by the turbocharger.

The "seemingly favorable" position for the engine curve referred to in the preceding paragraph is apparent in the figure: it passes through the region of highest compressor efficiency. Obtaining this position in matching turbocharger to engine was doubtless the engine designer's first objective; recall that the importance of high efficiency is brought out in the preceding section.

A second aspect of favorable position is avoidance of the compressor

96 SCAVENGING AND TURBOCHARGING

surge limit, the line in Figure 6.9 that forms the obvious left boundary of the characteristics plot. Beyond this boundary the contours of constant speed curve downward with decreasing flow, so that any momentary decrease in compressor delivery causes a decrease in flow, which causes a further decrease in pressure ratio. Flow may fall all the way to zero before it can resume at its steady value. Since the originating disturbance repeatedly occurs, this because of inevitable small randomness in the flow, a cycling between widely different flows is likely. This cycling upsets the operation of the engine, as well as being potentially damaging to the turbocharger. It is therefore to be avoided, and *is* avoided if the engine operating line stays well to the right of the boundary.

The engine line in Figure 6.9 is affected by the engine speed. Figure 6.10 is the same figure repeated, with several engine lines for lower speeds shown in addition. Also shown are contours of constant mean effective pressure. Since any path across the engine power-speed plane (such as that followed by a fixed-pitch propeller) is a locus of speed-MEP points, the path can be added to plots such as Figure 6.10. The engine designer's further task is to choose a turbocharger whose compressor surge line lies a safe distance from any such path that can occur.

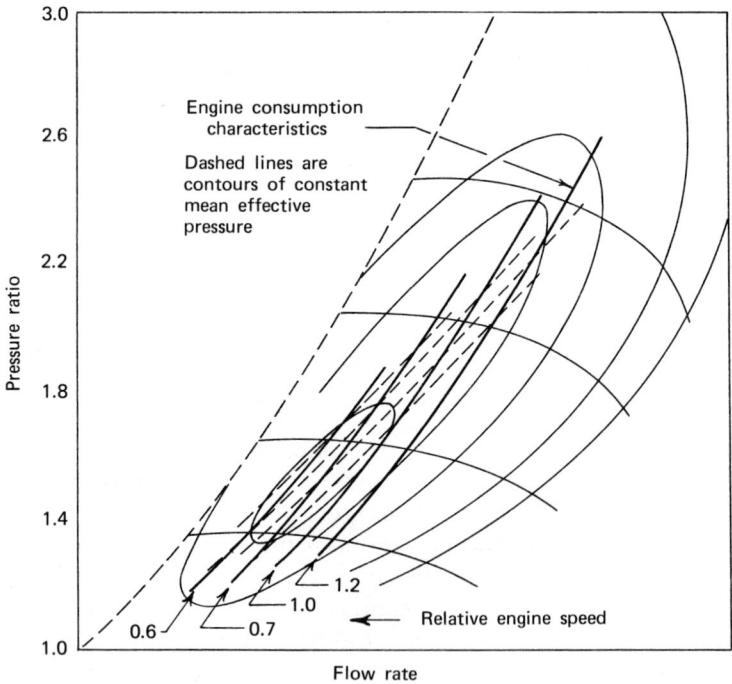

Figure 6.10 Figure 6.9 repeated with engine consumption characteristics added.

PULSE OPERATION AND CONSTANT PRESSURE OPERATION

Typical turbine characteristics are shown by Figure 6.11. The flow characteristic is essentially that of compressible flow through a nozzle; the influence of turbine speed, for instance, is very slight, and no such influence is included in the figure. Magnitude of flow for a particular pressure ratio is a function of nozzle area, a parameter that is always fixed for the turbines used in turbochargers.

Efficiency of the turbine is a strong function of its rotational speed, since it is principally a function of the speed ratio between nozzle jet speed and linear speed of the turbine blades. The effect of the relationship is evident in the figure.

6.5 PULSE OPERATION AND CONSTANT PRESSURE OPERATION

A previous section discusses the use of thermal energy from the engine cylinder to offset the irreversibilities of the turbocharger compression and expansion flows, an energy that is "free" when it is taken from the cylinder at a level unavailable to the piston. It further notes that at whatever level energy is taken from the cylinder, it may become inadequate at low powers, since cylinder temperatures are lower. A partial remedy for the energy deficiency is to conserve and use the kinetic energy in the exhaust gas as it flows from the cylinder—the "pulse" energy. To

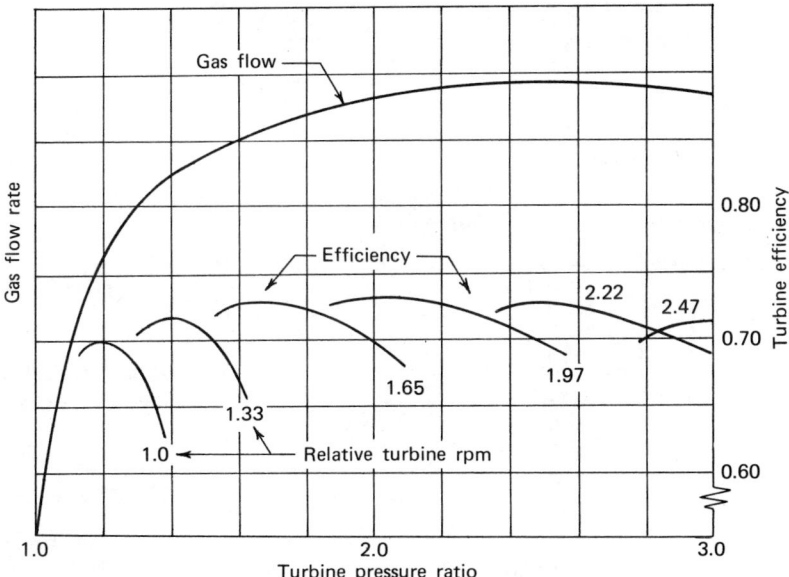

Figure 6.11 Typical turbocharger turbine characteristics.

do so, the engine designer must provide a flow path from exhaust port of valve to turbine nozzle that dissipates the least possible amount of fluid head. "Head" in this context refers to the kinetic energy; hence the flow path is short and with the minimum divergence in cross-sectional area.

An idea of the magnitude of the pulses at turbine inlet can be gained from the top sketch in Figure 6.12.

This pulse utilization has several disadvantages, however. One is the resulting poor efficiency of the turbine. For a particular rotational speed, a turbine has its peak efficiency at a particular nozzle jet speed. The pulses cause a widely varying jet speed, and varying at such a frequency that the turbine cannot accelerate and decelerate in response; turbine efficiency must inevitably be well below its maximum value.

Another disadvantage accompanying pulse utilization is the blow-back of exhaust gas from an exhausting cylinder into a neighbor, occurring because a peak from one cylinder will appear in the exhaust pipe while another is being scavenged at a lower pressure. For this reason, the best arrangement for pulse turbocharging has one unit for each group of three

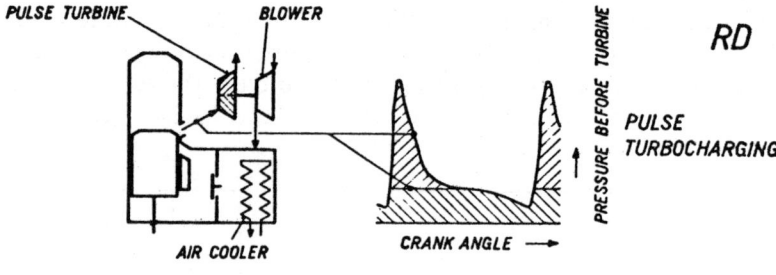

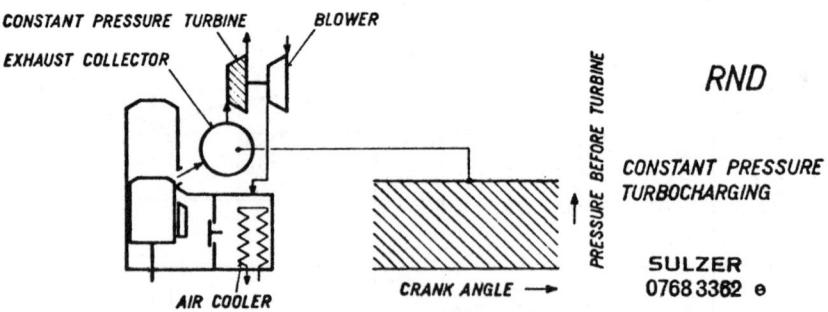

Figure 6.12 Some comparisons of pulse turbocharging and constant pressure turbocharging, Sulzer RD and RND engines [Sulzer (1968)].

PULSE OPERATION AND CONSTANT PRESSURE OPERATION

cylinders; with more than three, pulse overlap among cylinders connected to the same exhaust degrades the scavenging efficiency an intolerable amount; with fewer than three, too little pulse energy is supplied to the turbine. Sulzer pulse-charged engines, for example, are derated 5 to 10 percent when the number of cylinders is not an integral multiple of three [Sulzer Brothers (1968)].

The problem of exhaust blow-back usually requires manipulation of exhaust timing to reduce its magnitude. In engines with exhaust valves, timing of the valves is independent of inlet port opening, allowing adjustment of exhaust and scavenging events to minimize the problem. In engines with exhaust ports a rigid relationship among opening and closing events exists. For example, if exhaust ports open 20 crank angle degrees before the inlet ports open, then they must close 20 degrees after the inlet ports close. Auxiliary measures are feasible, however. Sulzer in its RD series engine, for example, uses a rotating valve in the exhaust line from each cylinder. This valve is closed during the sector of the piston cycle when exhaust ports are open and a pulse from another cylinder is present in the exhaust line.

The alternative to pulse operation is constant pressure operation. It is characterized by a large-volume exhaust receiver, quite in contrast to the direct velocity-conserving pipes of pulse charging. The kinetic energy of the pulses is deliberately degraded to heat by the sudden expansion of exhaust flow as it enters the receiver. Although less energy is thereby made available to the turbine, it is available at a constant pressure, resulting in a higher efficiency of its use. See Figure 6.12 for comparisons with several aspects of pulse operation. The blow-back problem is eliminated, so the the rule of three cylinders per turbocharger is not applicable; the number of turbochargers per engine consequently may be determined simply by the amount of air that available units can provide. As a result, a 12-cylinder engine using constant pressure charging may have two turbochargers in contrast to the four required with pulse charging. And other benfits are associated with constant pressure charging in some cases. For example, the rotary exhaust valves of the Sulzer RD engines are not found on the Sulzer RND series, since the latter uses the constant pressure alternative.

In spite of the several advantages of constant pressure charging, the need to make use of the pulse energy is overriding in engines at low-load conditions, and may obviate the need for mechanically driven blowers to provide charging air under these conditions. The engines of the first fifteen years or so of turbocharged low speed machines (roughly 1955-1970) therefore were either pulse charged, or used constant pressure charging only with major assistance from mechanically driven sources of air. As mean effective pressures, and hence scavenging pressures, were steadily raised, the balance tipped toward constant pressure with only minor help from other drivers. For example, the use of constant pressure charging by the Sulzer RND engines which appeared in the late 1960s has just been mentioned. MAN likewise produced its KSZ engines with constant pressure charging in the 1970s. The latter engines formerly used the piston underside as an auxiliary compressor at low loads, a feature that was replaced by a small electrically-driven blower when constant pressure charging was adopted.

6.6 TURBOCHARGERS COMBINED WITH MECHANICALLY DRIVEN BLOWERS

Air pumps to supplement the turbocharger are used by the low speed engines of several manufacturers. They are found in the form of electrically driven blowers used only at low loads or reciprocating units driven by linkages from the engine crossheads, or they may simply be the undersides of the engine pistons. These devices complicate the engine, of course, so that there has been some expectation thay they will be dispensed with as increasingly efficient turbochargers are able to use more of the exhaust energy. On the other hand, even when not essential they offer several attractions, namely:

1. In the event of turbocharger failure, they enable the engine to continue operation at part load.
2. If the turbocharger cannot provide the required air flow from energy unavailable to the pistons, an added mechanically driven blower is a more efficient solution than exhausting the cylinder earlier in the piston stroke. For example, Gyssler [Gyssler (1967)] states that providing 10 percent greater air flow will cost about 2.2 g/kWh in fuel consumption if a mechanically driven blower performs the service, but two to three times as much if it is done by exhausting earlier to increase turbine output.

For the reasons just cited, mechanical blowers appear (as of 1978) to be a permanent part of low speed engine technology.

The reciprocating compressors driven by linkages from the crossheads are exemplified by engines of Grand Motori Trieste (Fiat). Figure 6.13 is a cross-sectional view of a GMT engine with the compressor evident on the left side at the level of the crosshead. This unit is combined in a series flow arrangement with a constant pressure turbocharger in cross flow scavenging. The turbocharger compresses air from atmospheric to

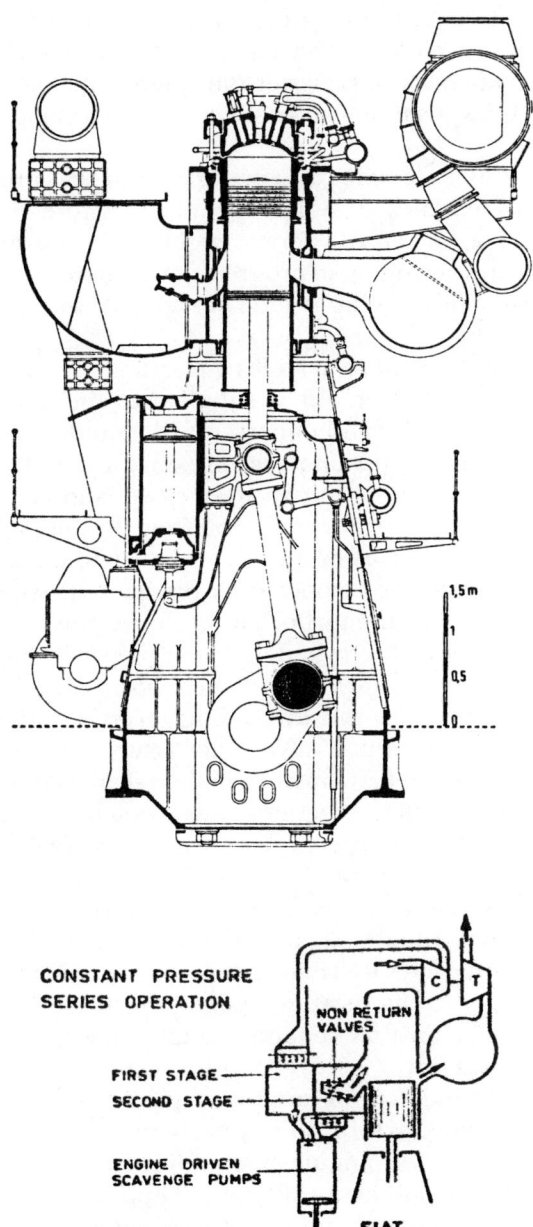

Figure 6.13 Cross section of Grand Motori Trieste (Fiat) engine showing reciprocating air pump [from Gyssler (1967)].

70 to 90 percent [Gyssler (1967)] of the scavenge pressure. The air passes through a cooler into a receiver which forms the suction chamber for the reciprocating compressor. The latter unit pumps the air through a second cooler into the scavenging air receiver. A sketch in Figure 6.13 shows these steps.

The use of piston undersides as scavenging and charging compressors is exemplified by engines of Sulzer Brothers. Figure 6.14 is a cross-sectional view of a Sulzer RD series engine. The piston is shown at bottom center with scavenging in progress; small arrows show the air flow paths at this time. This engine is loop scavenged, with pulse operation of the turbocharger.

Turbocharger and piston undersides are arranged in a combination of series and parallel by Sulzer. Air is supplied to the undersides by the turbocharger via the cooler and nonreturn valves indicated in the figure. As the down stroke of the piston raises pressure above the turbocharger discharge pressure, the nonreturn valves close, and compression procedes independently of the turbocharger. When the inlet ports are uncovered, the compressed air surges into the cylinder to initiate scavenging. Pressure quickly falls to the turbocharger discharge level, the nonreturn valves open, and air from the turbocharger thereupon flows through the scavenging receiver into the cylinder. It is this particular phase that is underway in Figure 6.14. After the cylinder ports close, flow continues from the turbocharger into the scavenging receiver and space beneath the pistons until the nonreturn valves once again close. The scavenging pressure when the ports first open is higher than that reached by the turbocharger alone—with consequent enhancement of scavenging—but only part of the air is compressed to this level, thus reducing the total energy used by the air system.

The later RND series of Sulzer uses constant pressure turbocharging, as noted earlier. The rotary exhaust valve of Figure 6.14 is not used; a similar cross section of the RND engine shows the exhaust receiver (one receiver per engine) in the location assigned to these valves in the RD series. Otherwise, the air flow sequence is the same as that described in the preceding paragraph.

Among the other builders of large low speed engines, Burmeister & Wain, Mitsubishi, and Doxford do not (in the late 1970s) use mechanically driven blowers. Their engines are uniflow scavenged with exhaust opening controlled by either a valve or an opposed piston (Doxford). Pulse operation of the turbocharger to provide sufficient air at low engine loads is used. MAN formerly employed piston underside compression in parallel with the turbocharger. As noted several times earlier, later MAN engines use no supplement to the turbocharger, save for a small motor-driven blower for very low loads.

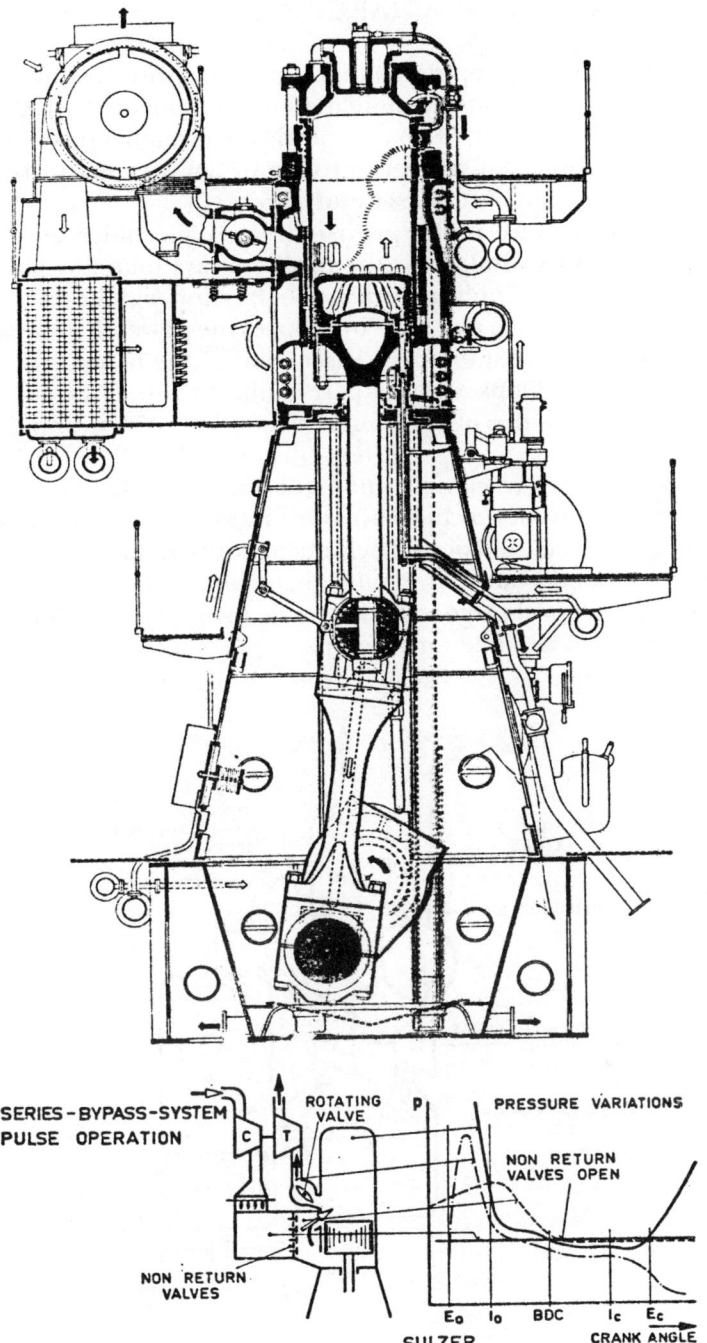

Figure 6.14 Cross section of Sulzer RD engine, showing scavenging in progress [from Gyssler (1967)].

6.7 TWO-STAGE TURBOCHARGING

An alternative to pulse operation and constant pressure operation of turbochargers is a two-stage combination of both. Figure 6.15 shows the arrangement schematically. Exhaust gas flows through a pulse-operated turbine, then via an exhaust receiver to a constant pressure turbine. The respective compressors are likewise in series.

An important feature is the provision of an air cooler between the compressors, this in addition to the cooler always found as part of the turbocharging outfit. Cooling during compression significantly reduces the work of compression, and thereby reduces the power that the turbines must deliver to the compressors. (Discussion of the benefits of cooling between stages of compression is part of the treatment of gas turbine power cycles in all thermodynamics texts [Van Wylen and Sonntag (1965)].)

The compressor characteristics of Figure 6.9 show the typical decline in compressor efficiency at high pressure ratios. Note that highest efficiency in the case pictured occurs below a pressure ratio of 2.0. When the pressure ratio being sought is greater than about 3.0, series connection of two compressors therefore allows each to operate nearer its most efficient zone than a single unit would.

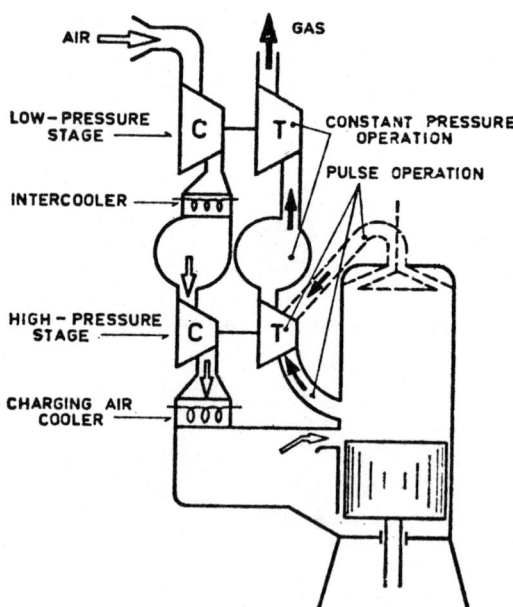

Figure 6.15 Arrangement for two-stage turbocharging [from Gyssler (1967)].

The saving in compressor work from the two factors just mentioned allows higher pressure ratios—hence higher engine mean effective pressures—to be reached without increasing the energy diverted to the turbocharger turbine. For this reason, two-stage turbocharging may be the key to continued improvement in the specific output (that is, increase in MEP) of low speed engines. Figure 6.16 shows the prediction of one authority [*THE MOTOR SHIP* (1973)] with respect to this possible benefit. It includes a historical record of the improvements gained with single stage charging since its introduction in the 1950s, but with a limit being reached at about 15 kPa; the predicted MEP with two-stage charging ranges upward into the neighborhood of 20 kPa.

The dashed curve in Figure 6.7 indicates the magnitude of an increase in pressure ratio attainable at a given turbocharger efficiency; compare this curve with the solid curve (single-stage) for the same efficiency.

The combination of pulse operation of the first turbine and constant pressure operation of the other allows the benefits of both to be obtained, since the distribution of work between the two shifts in a favorable way as engine load changes. The two units can be sized so that the low pressure compressor (the one driven by the constant pressure turbine) does most of the compression work (Gyssler [Gyssler (1967)] suggests 80 percent) at full engine load, thus gaining the benefit of the high efficiency

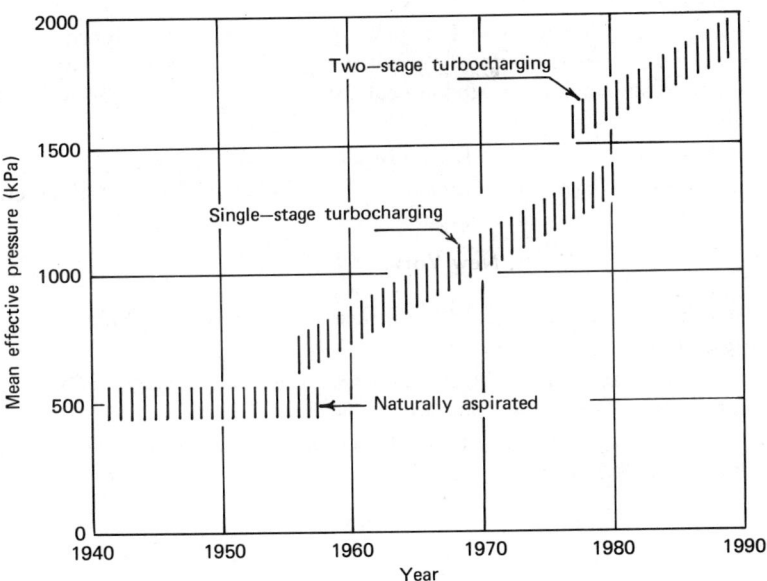

Figure 6.16 Mean effective pressure as a function of historical time [adapted from *THE MOTOR SHIP* (1973)].

of constant pressure operation when high pressure ratios are needed. As the load on the engine falls, the burden shifts naturally to the other unit, and it is in the low-load range that the greater availability of energy in the pulses is necessary for self-sustained turbocharger operation. With single-stage constant pressure charging, only uniflow scavenged engines with their valve control of exhaust can run at low loads without supplements to the turbocharger air flow. With two-stage charging, this benefit can be extended to engines using the other scavenging methods.

At the time this is written, only Mitsubishi offers a low speed engine with two-stage charging, although the advantages cited here have been recognized for many years. There is, quite naturally, a negative side, mainly that associated with the additional complexity, hence additional first cost, of the engine. Our illustrations (most recently Figures 6.13 and 6.14) show that the turbocharger and its associated coolers and ducting are components of considerable bulk; in addition, a turbocharger is a machine of precise design and manufacture, hence not cheap. The gain in specific output consequent to two-stage turbocharging is therefore to be balanced against a cost of significant magnitude.

6.8 REFERENCES

Brown Boveri Review (1977), Vol 64, 4.

Gyssler, G (1967), "Methods of Turbocharging with Special Reference to Large Two-Stroke Diesel Engines," *Journal of Engineering for Power*, Transactions of the American Society of Mechanical Engineers, Vol 89, 2, Series A, pages 71-86.

THE MOTOR SHIP Journal (1973), Vol 54, 641, page 430.

Obert, E F (1968), *Internal Combustion Engines—Analysis and Practice,* third edition, International Textbook Company.

Schweitzer, Paul H (1949), *Scavenging of Two-Stroke Diesel Engines,* The Macmillan Company, New York.

Sulzer Brothers Limited (1968), "Some New Developments in Sulzer Diesel Engines," unpublished notes.

Taylor, C F, and Taylor, E S (1961), *The Internal Combustion Engine,* second edition, International Textbook Company.

Van Wylen, G J, and Sonntag, R E (1965), *Fundamentals of Classical Thermodynamics,* John Wiley and Sons, New York.

6.9 NOTATION FOR CHAPTER 6

bar	bar (unit of pressure)
GMT	Grandi Motori Trieste
g/kWh	grams per kilowatt hour
kg/cm²	kilograms per square centimeter
kPa	kiloPascal
lb/in²	pounds force per square inch
MAN	Maschinenfabrik Augsburg Nürnburg
MEP	mean effective pressure
m_a	rate of air flow retained in cylinder
m_i	rate of air flow supplied to cylinder
N	rotational speed (rpm)
r	compression ratio
rpm	revolutions per minute
R_s	scavenging ratio
V_d	displacement volume
η_s	scavenging efficiency
ρ_s	density of air

Chapter Seven

ENGINE RATING

The *rating* of an engine is a number that expresses its capability. To introduce the concept in simple terms, one may consider the statement "the XXX-nnn engine is rated at 25,000 kW." This assertion is fairly explicit, seeming to mean that a ship it powers receives 25,000 kW to turn its propeller. But the concept is not really as simple as it seems from this brief statement; questions could be asked such as the following:

Under what conditions does it produce 25,000 kW?
Is this just a "rating," or is it the power actually produced?
At what shaft speed is this power produced?
Is it shaft power or brake power?

And a fifth question that really encompasses all of the others:

What must the marine designer know and do to establish the rating at this value?

The last question is included, and is important, because it is the designer of the ship's propulsion plant who sets the rating. True, the engine manufacturer is best acquainted with its product, and so publishes ratings, but the ship designer has the responsibility for choosing a service power—the power that the engine is actually to produce in the ship—and setting the rating accordingly. Note carefully—the manufacturer's rating may not be the rating established for the ship, and neither rating may coincide with the power actually produced. This chapter attempts to delineate the differences, and emphasizes the ship designer's responsibility in establishing their magnitudes.

7.1 FACTORS IN THE RATING

The number that most commonly designates the rating of an engine is its *power*. However, power applied to the output coupling consists of two factors, the torque and the rotational speed. Since these factors are nominally independent, a power rating must actually consist of two ratings: a rated *speed* and a rated *torque* (or its equivalent, *mean effective pressure*).

An engine obviously must be given an upper limit on operating speed. The reciprocating parts are continuously accelerating with the consequent inertia loads and stresses on these parts and their bearings. On the other hand, the undesired consequence of a moderately excessive speed is usually an unacceptable wear rate, not a catastrophic failure. For this reason, speeds that are not to be tolerated for continuous running may be used for brief periods without detectable harm to the engine. Engines are therefore often given an intermittent (or "trial trip") speed rating that is slightly higher than the continuous rating. Typically this intermittent rating is based on a power overload of 10 percent and hence is set at 103.2 percent of the continuous speed on the assumption that power as dictated by a propeller characteristic is proportional to cube of speed; note that 1.10 is the cube of 103.2.

An alternative scheme is to allow the 3 percent speed increase continuously if the power is not increased above 100 percent. A further 3 percent may be allowed for the trial trip. MAN, for example, follows this scheme. The 103 percent speed is expected to serve only when the ship is lightly loaded, as in the ballast condition, and engine governor is blocked to prevent this speed from being exceeded. On trials, a governor specialist temporarily removes the blockage, then replaces it when the high speed run is finished.

The builder of an engine most certainly will be better acquainted than the ship designer with the inertia loadings and stresses, and the consequent wear rates or other failure possibilities. The latter person will therefore doubtlessly accept as an upper limit that speed rating set by the builder, but may opt for a lower speed rating. For example, a rated speed of 110 rpm might be chosen in lieu of the builder's advertised continuous rating of 120, just for the purpose of obtaining a better propeller effi-

110 ENGINE RATING

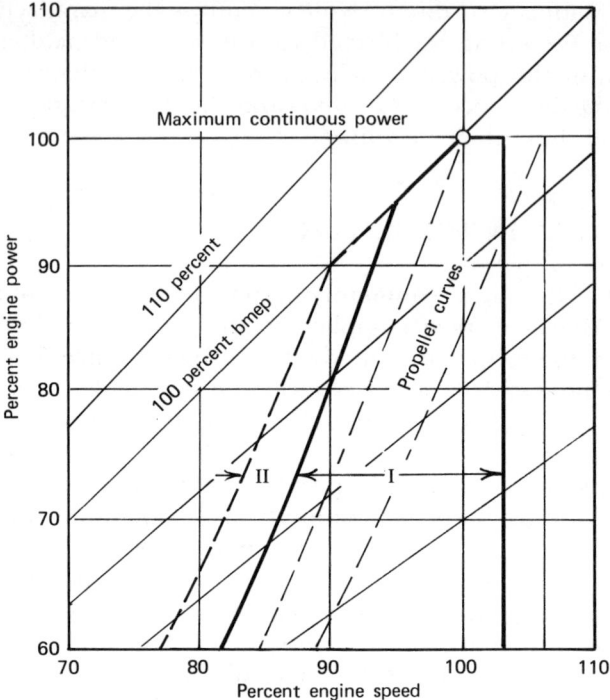

Figure 7.1 Typical engine builder's power–speed range for engine rating (adapted from MAN Figure 36 36 603).

ciency. However, a speed reduction means a power reduction, so that the ship designer is not likely to stray far from the maximum continuous rating published by the engine builder. Figure 7.1 illustrates a builder's expectation in this respect. It is based on a plot published by MAN (MAN figure 36 36 603) to indicate the speed ranges (as percentages) of its engines, and the corresponding powers allowed. Region I is the power-rpm domain acceptable for continuous operation, and you will note that it includes the continuous 3 percent overspeed concept discussed in the preceding paragraph, and the reduction of power at lower speeds. The Region II is the domain acceptable for transient operation.

The other factor of power rating, brake mean effective pressure* must also be given an upper limit, based on the amount of energy that can be released in a cylinder per cycle. The absolute limit is set by maximum

---

*We could speak of either torque or bmep here, since they are proportional, but the latter is more commonly used in discussions such as this.

capability of either fuel system or turbocharger to supply their respective ingredients. A somewhat lower limit is often set by acceptable levels of smoke in the exhaust, and a still lower level is set by the thermal loadings which cylinder components can endure over an acceptable lifetime. Just as in the case of speed, the engine builder is obviously best suited to establish a continuous bmep level, and so to set the maximum continuous power rating. The ship designer accepts this as an upper limit, and seeks the service power and the engine rating for the particular ship (the subject of Sections 7.3 through 7.6) within this limit.

Figure 7.1 indicates an obvious reduction in bmep required below about 95 percent speed. The reason is found among functions that are affected by engine speed. For example, cylinder lubrication is supplied by lubricators that work at the speed of the engine; the decrease in supply as speed decreases is therefore quite likely to require a reduction in bmep as well.

Before departing this section we should note several definitions. The first concerns the difference between *brake* power and *shaft* power. The former is the power at the engine output coupling, and is therefore the power published by the engine builder, as in Figure 7.1. Shaft power is the power delivered to the stern tube shaft coupling. Since low speed engines are nearly always directly connected to the shaft, no reduction gear losses intervene between engine and propeller, allowing shaft power to be only a trifling bit less than brake power. For this reason, the two powers are not distinguished in the discussions of this chapter, and brake power is implied wherever "power" is stated. However, shaft-attached machines such as electric power generators can be used, and then, of course, the designer must be alert to a major difference between brake and propeller powers; shaft power in this case is the sum of propeller power and generator power.

The rating most often stated by the engine builder is the *maximum continuous* brake power (often abbreviated MCR for "maximum continuous rating"), or the engine's power output at 100 percent bmep, 100 percent rpm as determined by the builder. This is the rating intended by the builder for continuous use under standard conditions, and is the starting point in the ship designer's determination of rating.

We should also note the definition of "maximum continuous" as stated by the Society of Naval Architects and Marine Engineers diesel bulletin [Society of Naval Architects and Marine Engineers (1975)]:

> Maximum continuous shaft horsepower is the shaft horsepower at which the power plant is designed to operate continuously and meet the requirements of the classification society. It may or may not coincide with the maximum continuous rating of the engine.

Now, in American practice the term "ABS power" or "ABS maximum power" often appears (ABS = American Bureau of Shipping). In steam practice, at least, this is the engine builder's maximum continuous power, the power used in any classification society formula (for propulsion shaft diameter, for example), and the ship designer's rating as *the* power installed in the ship. Everyone agrees, and consequently if speaking of steam, that "may or may not coincide" caveat would not appear. In diesel practice, certain margins are necessary from the standpoint of the ship designer, and it is these that cause the uncertainty in the definition here. "Maximum continuous rating of the engine" in the definition coincides with the diesel builder's maximum continous rating. The other power may be less by the margins mentioned; what these margins may be is to be discussed in Sections 7.4 and 7.5.

7.2 RATING CORRECTIONS

The first step among the several that lie between engine builder's maximum continous rating and the ship's rating and service power is the adjustment in the builder's rating that the particular ship environment may require.

The environment affects the output of a diesel engine in a number of ways. The effect of temperature of inlet air on exhaust temperature, for instance, has been discussed in Chapter 4. The density of atmospheric air affects performance because it influences the mass of air trapped in the cylinder when the intake ports close, and this density is a function of atmospheric pressure, temperature, and humidity. Upon reaching the cylinder, however, the air will have been modified in its state by the turbocharger compressor, and by the intercooler between compressor and cylinder. The performance of both of these units can be affected by ambient conditions—the compressor by inlet density, and the cooler by temperature of the water supplied to it. For these reasons, the rating of an engine must be specified under standard conditions, and subsequently modified if service conditions are significantly different from the standard. For instance, design of a ship that will operate in the tropics may require such attention.

An engine manufacturer may specify standard conditions that are appropriate to its product, or may use the conditions published by a technical society or standards-setting organization. In the USA, for example, many diesel manufacturers rate their engines by the standards of the Society of Automotive Engineers. The European builders of low speed marine diesels may use the standards of the International Standards Organization (ISO). However, their practices may change, and usually show some differences. For example, in 1979 MAN used 45 C air temperature in rating its KSZ engines, Sulzer used 27 C in rating its RLA and RND engines, and Burmeister & Wain used 20 C in rating its K and L engines [Bohm and Simon (1979)].

Presumably service conditions will be no worse (temperatures no higher) than the standard used by the builder, but nonetheless it is a duty of the ship designer to check this assumption, and if it be wrong, to make corrections ("deratings") to suit. Doing so in a completely rational way is difficult because of the differing responses of turbocharger and of cylinder to changes in ambient conditions; formulas that are used tend in consequence to be of the rule-of-thumb variety. And rule-of-thumb is all we shall attempt here, and I quote the formulas offered by the Society of Naval Architects and Marine Engineers bulletin [Society of Naval Architects and Marine Engineers (1975)]:

Derate the engine 10 percent for each 4 inch Hg reduction in barometric pressure.

Derate the engine 2.5 percent for each 10 F increase in ambient air temperature.

Derate the engine 1 percent for each 10 percent increase in relative humidity.

For intercooled engines if sea water is used, derate the engine 2 percent for each 10 F increase in sea water temperature.

These formulas are said by the bulletin to be for preliminary design when better data are not available. As a design proceeds beyond the preliminary stage, any necessary deratings might be improved by asking the engine builder to furnish correction formulas appropriate to its product.

A student wishing to pursue the derating question to its theoretical foundations can find a technical literature of modest extent. An example has reached my desk at the time of this writing is a Society of Automotive Engineers paper by Wu and McAulay [Wu and McAulay (1973)]. Its list of references leads to other sources.

The "back pressure" on an engine—the pressure at engine exhaust point that is required to send the exhaust gas to atmosphere—also affects engine performance, and hence its rating. The usual design situation is that the engine builder specifies the back pressure for which the published rating applies, and the ship designer works to this figure in laying out the exhaust line from engine to atmosphere. But if circumstances dictate an exhaust system that requires excessive back pressure, then a derating is required. Again we quote the Society of Naval Architects and Marine Engineers bulletin [Society of Naval Architects and Marine Engi-

114 ENGINE RATING

neers (1975)] for preliminary design use, and advise later use of data furnished by the builder. The quotation follows:

> Derate the engine 1 percent for each 4 inches of water increase in exhaust back pressure.

7.3 CHOOSING RPM

The power that an engine will produce depends on the power-speed characteristic of the propeller, a message that has been elaborated in Chapter 5. The ship designer thus effectively sets the engine speed by choice of propeller, and is expected to choose it so that the service speed of the ship occurs at the 100 percent rpm of the engine builder's maximum continuous rating. But in seeking better propeller efficiency (for example), the designer may choose a lower speed. Recall, though, that a speed reduction requires a power reduction, most recently illustrated by Figure 7.1. Although noting the possibility here of speed choice as a possible step in rating selection, the following sections are written on the assumption that the ship designer's 100 percent rpm will be the same as the engine builder's 100 percent rpm.

7.4 OPERATING MARGIN

The most nettlesome rating question that faces the ship designer is that of "operating margin": should the engine be derated (bmep reduced below the builder's maximum continuous value) to improve its expected maintenance history through lighter mechanical and thermal loads? The engine builder is likely to say no, this on the argument that the maximum continuous rating means just what it implies. A conservative shipowner, perhaps carrying unpleasant memories of past maintenance difficulties, may reply quite affirmatively.

If a margin *is* to be chosen, either through the designer's judgment or owner's specification, what magnitude should it have? There's no easy answer to this; I can only say that a commonly cited figure [Bullock (1979)] is 10 percent, and that I can't recall hearing a figure greater than 15 percent.

7.5 HULL SERVICE MARGIN

Resistance and powering design analyses are usually made for ideal conditions: smooth surface of hull and propeller, and calm wind and water.

HULL SERVICE MARGIN   115

Departures from ideal inevitably occur in service, so that margins are added to allow design speed under average service conditions. For example, a specification often applied in American practice requires that a ship produce its design speed on trial trip using 80 percent of its rated power [Maritime Administration (1962)].

Recall from Chapter 5 that a margin of this type is almost essential for a diesel engine driving a fixed-pitch propeller, since any increase in hull resistance causes a reduction in power unless the bmep can be increased to compensate.

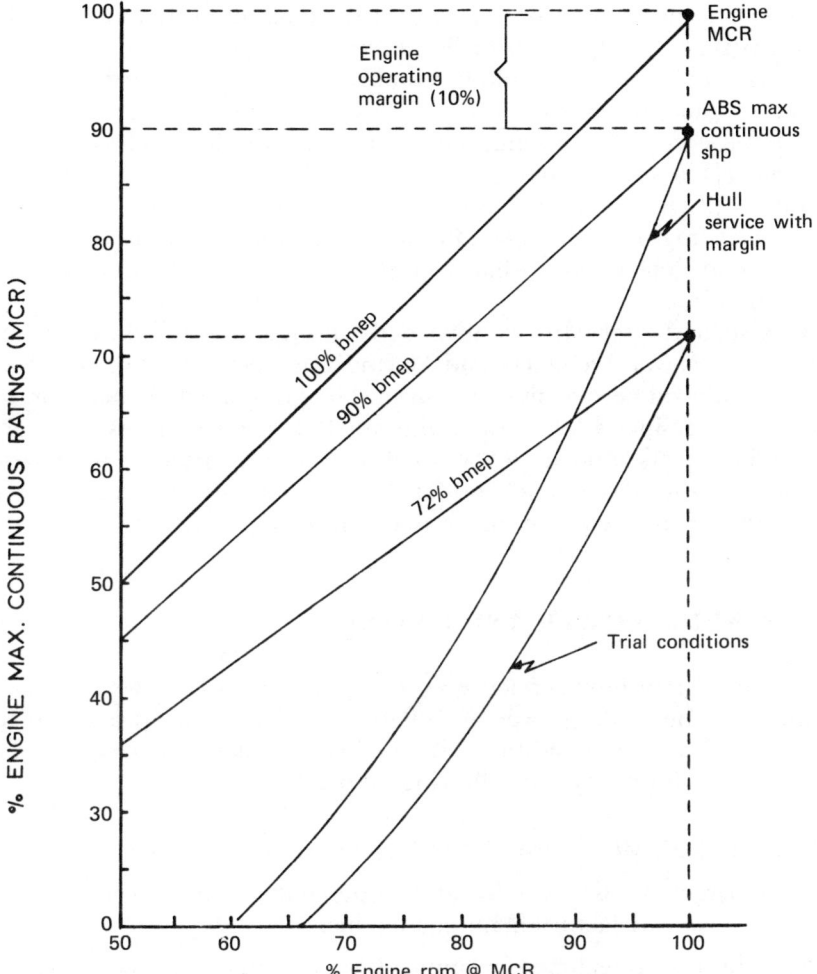

Figure 7.2 Illustration of margins on the power vs rPM plane [adapted from Bullock (1979)].

116 ENGINE RATING

Figure 7.2 illustrates the rating situation if a 10 percent operating margin is applied, and a hull service margin of the magnitude just mentioned is subsequently applied. Note that on trials, only 72 percent of engine builder's maximum continuous power can be developed unless the temporary speed increase mentioned in Section 7.1 is allowed. In service, "full power" would vary along the 100 percent rpm line between 72 and 90 percent bmep as weather and hull roughness changed.

A question now suggests itself: does Figure 7.2 represent excessive application of margin? The 10 percent operating margin apparently was thought by the hypothetical designer here to be adequate for a good maintenance history, yet the actual margin is much greater because weather and hull conditions will rarely demand all of that additional 20 percent (that is, 20 percent of 90 percent) margin.

Perhaps the margins should be combined, simply by selecting the propeller so that the *average* power is x percent of the engine builder's maximum continuous rating. The value of x would be chosen by two criteria: (1) it is low enough to provide the maintenance history demanded; (2) it is high enough to provide the speed specified under service (average) conditions. The principal consequent task for the designer is in determining what average conditions are in terms of hull resistance.

This second alternative is apparently intended by the diesel bulletin [Society of Naval Architects and Marine Engineers (1975)]; see Figure 7.3, reproduced here from that source. The point labeled "max continuous" is the builder's MCR; the point labeled "service hp" is a power to be used on trials, a power low enough that average power will be below maximum continuous, and enough below to satisfy the designer's requirements for acceptable maintenance history.

7.6 SERVICE POWER AND RATING

Let's summarize how service power is obtained, where *service power* is defined as the brake power at 100 percent rpm, and in average hull surface and weather conditions. By the first alternative discussed in Section 7.5 it is found by the following steps:

1. Builder's MCR is corrected for environmental factors.
2. Corrected MCR is reduced by a percentage (10 percent or other) to give satisfactory maintenance history.
3. The resulting bmep is reduced by an additional percentage (20 percent or other) to provide hull margin.

SERVICE POWER AND RATING 117

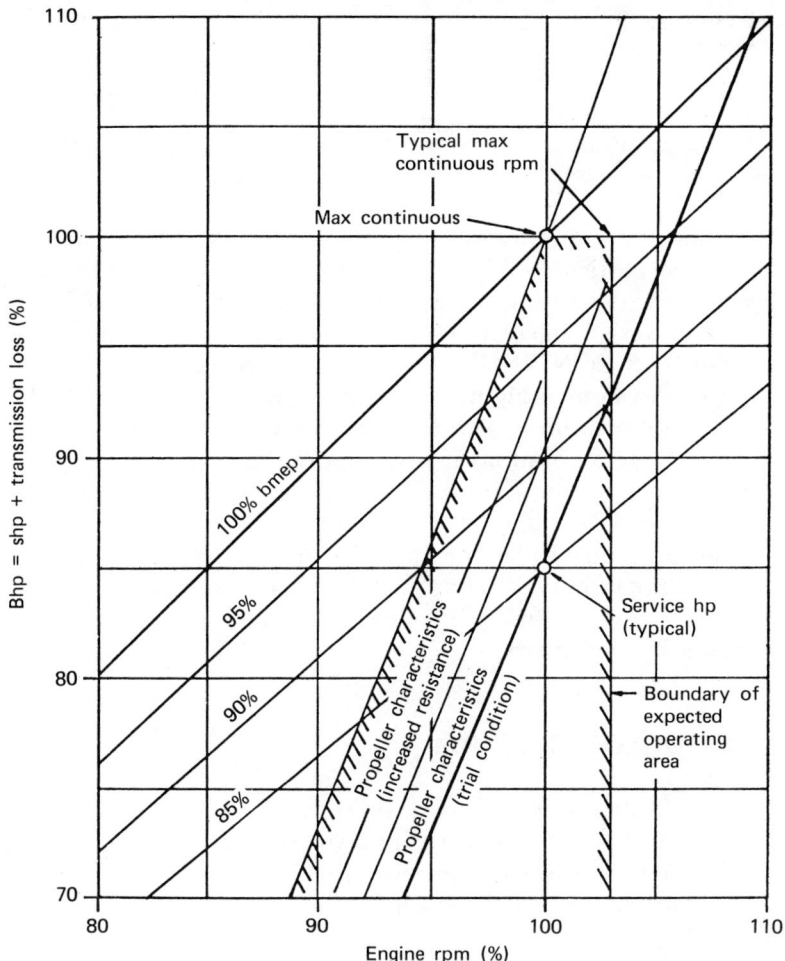

Figure 7.3 Relationship of service power and maximum continuous rating according to Naval Architects and Marine Engineers diesel bulletin [Society of Naval Architects and Marine Engineers (1975), by permission].

By the second alternative discussed in Section 7.5 it is found by the following steps:

1. Builder's MCR is corrected for environmental factors.
2. Corrected MCR is reduced by a percentage (10 percent or other) to give satisfactory maintenance history.
3. A trial trip bmep is chosen (and obtained via propeller selection)

118 ENGINE RATING

so that the power resulting from the preceding step occurs under average conditions.

In the second alternative, unlike the first, the value of service power has a definite value, although actual power obviously varies just as much as conditions change. But the service power, however found, is *not* the *rating*.

Let's define the *rating* as the maximum power that the engine can produce in the particular ship. This is the power that should be published, that coincides with the "ABS power" mentioned previously, and that is used in design calculations (for example, in finding shaft diameter). It is the MAX CONTINUOUS point in Figure 7.3, or the ABS MAX CONTINUOUS point in Figure 7.2. It may coincide with the builder's maximum continous rating, as it appears to do in Figure 7.3, or may be lower by a definite amount, as it is in Figure 7.2.

"Power that the engine can produce" depends on two factors: (1) the propeller characteristic, and (2) the limitations on rpm and bmep imposed by the governor. In Figure 7.3, for example, the governor presumably sets the limits implied by the crosshatched boundaries. The rating is therefore the power at the point where the highest possible propeller curve crosses this boundary.

7.7 COMPARISON CRITERIA

The process of establishing a service power as outlined in the preceding sections is predicated on accepting the manufacturer's maximum continuous rating as a fixed starting point for reaching a rating through one or more downward steps. This is based on the notion that the reader is a student or novice designer, and that such a person will have to respect the manufacturer's superior knowledge of its own product. A designer is nonetheless expected to gain in knowledge through continuing experience in choosing ratings and service powers, in digesting the consequent feedback, and generally through continuously increasing knowledge of the engines. The experienced designer may never reach the point of independence from the builder's data, but may be able to judge the degrees of optimism or conservatism used by manufacturers of engines, and make good use of this judgment in altering published ratings downward into service values, as in setting the value of the operating margin.

All of the engine builders are doubtless honest in rating their engines —they have to be—but they are under competitive pressure to rate them highly, for the purchaser naturally wants the most power per dollar and per kilogram of engine weight. Given, then, that different people and different organizations work under differing philosophies and different levels of optimism and conservatism, it is reasonable to suggest that a ship designer be capable of making independent judgments on ratings.

Judgments must be based on quantitive information—on data that relate to loadings and stresses, such as piston speed, mean effective pressure, maximum cylinder pressure, etc. The major purpose of this section is to point out several parameters that can serve as judgment bases. The first of these is the *specific output* (or *specific power*), which is simply the power per unit piston area (kW/A, bhp/A). Recall from Chapter 2 that power is the product of P, L, A, N (mep, stroke length, piston area, rotational speed) and that piston speed is proportional to the product of L and N. The specific output must be PLAN divided by A and hence is proportional to the product of mep and piston speed. It therefore encompasses a parameter that is related to thermal loadings, and one that is related to inertia loadings. Specific output, easy to evaluate from the usual published data of engine builders, serves readily to compare engines; if engine A has a higher specific output than engine B, then there is ground for asking "is A rated less conservatively (hence more subject to wear) than B?" It may be, of course, that A is designed to a higher technical level than B (and may cost more in consequence), and so can sustain a higher specific output. For such a reason are deeper knowledge and experience needed in the judgments; for such reasons no categorical statement can be made here about acceptable and unacceptable levels of bhp/A or kW/A.

The curve of Figure 7.4 illustrates another approach [Wagner (1970)]. It is a plot of estimated engine life between overhauls as a function of a cylinder loading parameter, the latter being F/LA, where F is the fuel consumed per unit time, and LA is the cylinder displacement. The curve was developed from experiences of several North American builders of medium and high speed engines. Because the results cannot therefore be directly applicable to low speed engines, and because these results (overhaul lives) are likely to be obsolete before the curve is published here, numerical scales are omitted. If one may assume that overhaul lives of low speed engines are affected by the same factors that affect the lives of medium and high speed engines, then the abscissa parameter can nonetheless be used as a comparative judgment parameter in the same fashion as specific output.

A parameter called *differential thermal efficiency* was advocated by Cramer and Froelich [Cramer and Froelich (1956)] as a means of judging ratings. It is based on the notion that increasing amounts of fuel per cycle produce increasing amounts of output, but that the ratio (output increment)/(fuel increment) decreases ("diminishing returns"). In the limit, the cylinder is unable to burn any more fuel, and the ratio has

120 ENGINE RATING

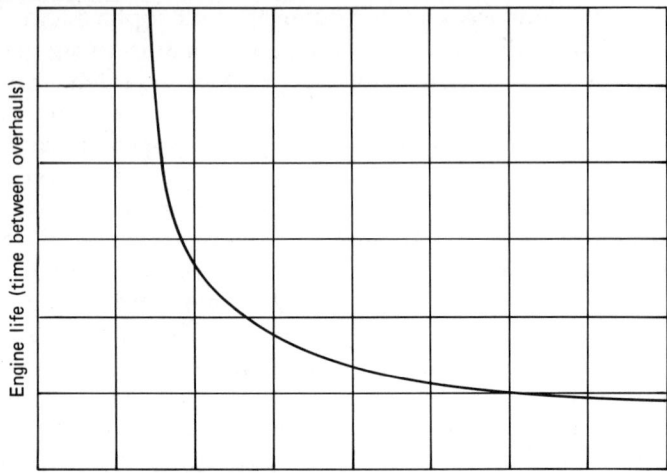

Figure 7.4 Predicted engine life as a function of a fuel input parameter [adapted from Wagner (1970)].

become zero. Some point well below this extreme is appropriate for the rating, and presumably all engines rated with equal conservatism should show about the same value of the ratio at their rated points. The concept is shown graphically by Figure 7.5, this being a plot of output in terms of bmep as a function of a parameter related to fuel input, a parameter whose formula is

$$\text{Fuel parameter} = \frac{F\;HV}{LAN} \tag{7.1}$$

where F is fuel consumption in mass per unit time, and HV is the heating value of the fuel. The slope of this curve is the differential thermal efficiency, and is plotted as a second curve in Figure 7.5.

The task of a ship designer would be simplified if we could state values for any of the three parameters introduced here, and say "this is an appropriate value -- beware of an engine with a higher value." Such assertions are impossible because, first, changing technology may allow ever higher ratings, and second, because all of these parameters are oversimplifications—they obviously do not give any notion of the many subtleties of engine technology. In consequence they serve only as starting points, or as tools for organizing experience. A designer may note that the specific output or differential thermal efficiency (for examples) of engine A is higher than that of engine B, but this is not an answer, but only a pointer to further consideration. To close by repeating an earlier message, the difference may not be unconservative rating practice by the

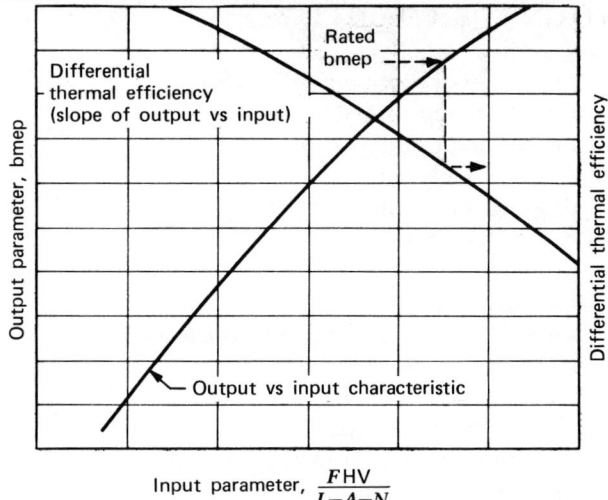

Figure 7.5 Differential thermal efficiency as a function of a fuel input parameter [adapted from Cramer and Froelich (1956)].

builder of A, but a technical superiority that makes the higher level appropriate.

7.8 REFERENCES

Bohm, F and Simon, J (1979), "The Fuel Consumption Rates of Present-Day Two-Stroke Engines," published and distributed by MAN.

Bullock, W G (1979), "Power and RPM Determination in the Selection of Diesel Engines for Merchant Ship Main Propulsion," a letter to selected authorities, Maritime Administration, U S Department of Commerce.

Cramer, R J and Froelich, K F (1956), "Rating Engines by Fuel-Air Ratio Effects," American Society of Mechanical Engineers, Oil and Gas Power Proceedings.

Maritime Administration, U S Department of Commerce (1962), Design Memo 19, Office of Ship Construction.

Society of Naval Architects and Marine Engineers (1975), *Marine Diesel Power Plant Performance Practices,* Technical and Research Bulletin 3-27.

Wagner, Michael (1970), "Rational Methods for Engine Rating," University of Michigan Department of Naval Architecture and Marine Engineering, Report 090.

Wu, T and McAulay, K J (1973), "Predicting Diesel Engine Performance at Various Ambient Conditions," *Transactions,* Society of Automotive Engineers, paper 730148.

7.9 NOTATION FOR CHAPTER 7

A	piston face area
bhp	brake horsepower
bmep	brake mean effective pressure
C	degrees Celsius
F	degrees Fahrenheit
F	fuel consumption per unit time
Hg	mercury
HV	fuel heating value
ISO	international standards organization
kcal/kg	kilocalories per kilogram
kW	kilowatt
L	piston stroke length
MAN	Maschinenfabrik Augsburg Nurnburg
MCR	maximum continuous rating
mep	mean effective pressure
N	engine speed
P	mean effective pressure
rpm	rotational speed (revolutions per minute)
shp	shaft horsepower

Chapter

Eight

FUELS AND LUBRICANTS

As in all types of marine engines, the principal fuels for the low speed diesel are petroleum-based liquids. This chapter discusses the properties of fuels that fall within this category, the significance of these properties to the engine, and a few things that may be done when the properties are harmful. The only likely alternative to the petroleum liquids is gas, specifically "boil-off" from liquefied natural gas cargoes, and a discussion of this fuel is also included.*

Fuel oils have many properties of significance, but an introductory discussion can be based on just two. They are the viscosity and the impurity content. Viscosity is the property of greatest significance in fixing the pumping and atomizing behavior of the fuel, and is the most commonly used in simple classification of fuels: a "heavy" fuel is one that has a high viscosity at ordinary temperatures, while a "light" fuel has a low viscosity. Impurities of several kinds are troublesome because of clogging of filters and injectors, deposition of their combustion products within the engine and turbocharger, and the accelerated wear of engine hot parts that these products may promote.

Fuel properties can be improved to some extent to suit the requirements of the engine. Viscosity is readily reduced by heating, and some

---

*Oil-coal slurries, and possibly coal itself, seem possible, but not in the immediate future at the date of this book.

123

impurities can be removed, or can be neutralized by additives or by the action of a neutralizing lubricating oil. Fuels used by marine low speed engines are most commonly of poor quality, so that these improvement processes are an essential part of their fuel systems.

The petroleum fuels used widely in marine service may be classified as either *medium distillates* or as *residuals*. These terms refer to the key characteristics of the refining process that produces them. Refining is fundamentally a process of distillation by which the many constituents of crude petroleum are separated according to their unique boiling points into individual products, such as gasolines or fuel oils. Any product produced wholly in this way many be called a distillate; the residue of the refining process—the constituents that have not distilled because their boiling points are impractically high—are the residuals.

A "medium" distillate is one whose temperature range of distillation is intermediate among the products normally used as fuels. Two such distillates have long been available from marine fuel suppliers for commercial diesel use; they are commonly known as *marine gas oil* and *marine diesel oil.* The gas oil is the lighter of the two (that is, lower distillation temperature, lower viscosity) and is used by high speed diesel engines. The other is suitable for medium speed and low speed engines, and can be used by them without heating or other treatment.

Residual fuel oil (often called "bunker C") is the highly viscous remnant of the refining process, and is the traditional boiler fuel for commercial steamships. Heating is always required to reduce its viscosity for handling and atomization, but steam boilers usually burn it without other treatment, even though it retains most of the impurities that were in the original crude oil.

A range of marine fuels known as *intermediates* is prepared by blending distillate with residual, the purpose being to produce a low-price fuel that is of acceptable quality for low speed and medium speed diesels whose owners do not wish to risk the uncertain properties of undiluted residuals.

A partial alternative to residual oil as a cheap fuel is a *heavy distillate,* that is, a fuel of higher viscosity than marine diesel oil, but one containing none of the residual component found in the traditional intermediates. The basic idea is that impurities should not distill, but remain in the residue. This is an ideal not completely realized, and in fact some impurities may originate in the refining process and in subsequent handling, but nonetheless, impurities are sharply reduced when residual oil is kept out of the fuel. The heavy distillate is a relatively new fuel in marine experience, its specifications having been developed in the late 1960s as a consequence of the U. S. Navy's need to have a single fuel that could serve boilers, diesels, and gas turbines alike. But in the commercial marine field the success of the low speed diesel in burning fuels that are largely residual has blunted the appeal of the heavy distillate concept.

Lubricating oil is also discussed in this chapter. Generally speaking, lubricants and fuels are two quite different things, save for their common origin as petroleum-based liquids. In the case of the low speed diesel, however, the properties of the oil used to lubricate the cylinders are strongly influenced by the properties of the fuel burned there. The contaminants in low quality fuel expedite corrosion-wear to such an extent that this fuel is probably unusable unless postcombustion neutralization of contaminant products is accomplished. Special cylinder lubricating oils, with properties designed to counter those of the fuel, do accomplish this, and thus are a key factor in the economic success of the low speed engine.

8.1 FUEL PROPERTIES

Each of the several fuels has properties that determine its behavior in combustion, its safety, its ease of handling, and its aftereffects (corrosive deposits, for example) on the machinery using it. Some of them are of interest over a wide range of applications, and so are of general concern; others are of concern only in a particular application. The discussion in this section begins with the general concern, then in its third sub section moves into the particulars of low speed diesel fuel.

A. Physical Properties

Viscosity is that property that determines the resistance to flow of a fluid. Because of its importance to fuel technology, and because a number of systems of units are used for quantifying it, an appendix appears at the end of this chapter in which a more precise definition is given, and in which several common systems of measurement are explained.

High viscosity makes pumping difficult, and makes atomization for combustion even more difficult, perhaps impossible. Fortunately, viscosity of liquids is a strong inverse function of temperature, so that moderate heating usually suffices to reduce viscosity of all common marine fuels to acceptable values, whatever the type of engine.

A low viscosity may be troublesome only if it is used in a piping system designed for a higher viscosity liquid. Seals that are adequate for the more viscous liquid may leak, and pumps that should be lubricated by the fluid pumped may not be so well served by the lower viscosity liquid.

126 FUELS AND LUBRICANTS

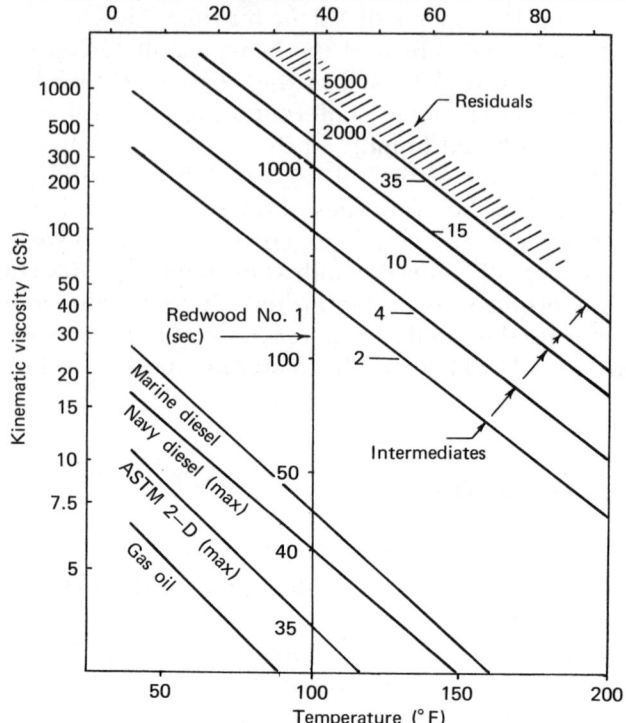

Figure 8.1 Viscosities of some marine fuel oils.

Figure 8.1 shows the viscosities expected for several of the fuels discussed in this chapter.

Distillation temperatures are usually specified by the temperatures at which 10 percent, 90 percent, and 100 percent of the product (end point) will have evaporated from the crude oil sources. These temperatures are of no direct concern to the engine using the fuel, but the temperature range essentially fixes the list of compounds that will appear in the distilled product, and so indirectly fixes all properties. Burning properties, in particular, are closely related to distillation temperatures. A low 10 percent temperature indicates the presence of highly volatile compounds, making the fuel easy to ignite; a high 90 or 100 percent value indicates the presence of long-burning compounds that in a boiler or gas turbine may produce an unacceptably long flame.

Flash point is mainly a safety criterion, being the temperature at which fuel will ignite momentarily, but will not continue to burn when the source of ignition is removed. It is a low flash point that distinguishes gasoline as a hazardous fuel for marine use.

Pour point is the temperature at which fuel just begins to flow, and so is the property that determines handling characteristics under cold conditions. It is obviously closely related to viscosity. *Cloud point* is also a low-temperature characteristic, being the temperature at which a waxy constituent begins to congeal in the fuel. This constituent can clog filters when filtration of a cold fuel is attempted. Heating is the usual remedy for unsatisfactory pour point and cloud point.

Carbon residue percentage is an indication of the amount of unburned carbon that may appear in the combustion process. It is typically quite low in all fuels except the residual oil. In that fuel it is often identified with semisolid asphaltene compounds that can be removed by centrifuging or other mechanical process.

Heating value is that amount of thermal energy released by the products of combustion of a standard amount of fuel, when those products are cooled to a base temperature. *Higher* heating value is obtained if this cooling process condenses the water vapor that originates from the combustion of the hydrogen fraction of the fuel; *lower* heating value is obtained if this condensation does not take place. The importance of heating value as a fuel property should be obvious. After all, it's what the user is paying for.

B. Contaminant Properties

Contaminants may be normal constituents of the fuel, that is, minor elements or compounds present in the crude oil, or they may be things such as water and rust scale picked up in handling and storage. In the former category are substances containing perhaps several dozen elements in detectable quantity. Those of significance to marine power plants are more limited in number; those usually regarded as significant are sulfur and vanadium. Other metals, such as sodium, potassium, calcium, lead, and copper, are important in some uses (particularly gas turbines) if present in sufficient concentrations. Although the items on this latter list usually do not appear to significant degree in petroleum, contamination from handling and storage often provides them. Sodium, for example, is abundantly provided by sodium chloride if the fuel becomes contaminated with seawater.

Sulfur is usually the element of highest concentration in crude petroleum after the carbon and hydrogen. Most of it ends up in residual fuel, where it may be 2 to 5 percent by weight, but smaller percentages also appear in distillates. It has been identified as a deleterious component of fuel in marine propulsion plants of all oil-burning types, especially when heat-recovery equipment (waste heat boiler with diesel and gas turbine, boiler air heater, or economizer with steam) is used, since the resulting low exhaust gas temperature can bring the gas below its dew point. Water condensing from an exhaust gas produced by a sulfurous fuel is in effect a dilute solution of sulfuric acid, and consequently metal surfaces are rapidly corroded.

Other solid contaminants of noncombustible nature are often referred to as *ash,* and are principally metals or metallic compounds. The engine most sensitive to fuel contaminants is the gas turbine; for it the ash components usually regarded as harmful are vanadium, calcium, sodium, potassium, copper, and lead. Vanadium, and sodium if present with vanadium, are also of concern to steam boilers and to diesel engines. As noted, only vanadium among those listed in this paragraph originates in the crude oil. Lead, for example, may be present if the fuel has been stored in a tank that previously held leaded gasoline. Lead and copper both may be contaminants in the refining process. Sodium, potassium, and calcium most commonly originate from seawater contamination, and are likely to occur if fuel is stored in tanks used alternatively as ballast tanks, either aboard the using ship, or aboard a vessel that transported it from the refinery to the fuel depot.

Whatever the engine type, vanadium is usually regarded as the most serious of these "ash" contaminants, since it is the key constituent in a deleterious vanadium-sodium (or potassium)-oxygen compound that is formed in the combustion process. This compound is liquid at temperatures of about 500 C and above. In this state it adheres to hot metal surfaces, and subsequently may corrode them.

Water is usually present in fuel in small fractions that, if truly small, are not harmful as such. But there are several secondary effects. One has already been mentioned, namely, the contribution of metallic contaminants when the water comes from the sea. Another is its service as a culture medium for bacteriological growth; certain microorganisms feed on hydrocarbon compounds that may have dissolved in water from the fuel, or are present at oil-water interfaces. The product of this activity is slime that clogs fuel filters, not only restricting fuel flow, but reducing the effectiveness of the filter in removing other solid matter. The problem usually occurs only with fuel that has been stored a long time, since the growth process proceeds slowly under normal circumstances. When conditions are favorable for this phenomenon, biocides can be added to inhibit the growth of the bacteria.

Sediment in fuel or particulate matter—dirt, to give it a plain name—is almost certain to be present, having originated in a number of different ways. Scale from the surfaces of tanks and pipes is an obvious source for gritty stuff, as is dirt remaining from the building of the tanks. And the sediment produced by biological activity has just been mentioned. It is for these stray solid materials that filters are universally required in fuel systems.

C. Diesel Properties

The properties discussed to this point are of concern to users of oil fuels in general. For diesel engine use, an additional property, that of *ignition quality*, is important. Reflect that the combustion of a diesel fuel takes place in batch fashion, with each combustion event perforce occurring within a small fraction of a second. A short ignition lag, or time from beginning of injection to beginning of combustion, is therefore essential, and this lag is the only measurable element of "quality."

Ignition lag and hence ignition quality, is denoted by a *cetane number*, a parameter whose value lies on a scale of 15 to 100; a high value signifies high ignition quality by giving a measure of how closely the fuel approaches the behavior of the reference compound, n-cetane. This hydrocarbon compound has the shortest lag among those likely to appear in a liquid fuel, and hence is suitable as a reference by which all fuels can be judged. The reference compound for poor quality is heptamethylnonane (HMN), and is given the arbitrary cetane number of 15, thereby establishing the lower end of the ignition quality scale.

To establish the cetane number of a fuel, it is burned in a standard test engine instrumented for the measurement of the time interval between beginning of injection and the sharp pressure rise that signifies combustion. Reference fuels consisting of mixtures of n-cetane and HMN are similarly run. These reference fuels have cetane numbers defined by (volume percent cetane + 0.15 × volume percent HMN); the tested fuel then takes its cetane number from the reference fuel that records the same lag. The test is described in detail by ASTM (American Society for Testing and Materials) standard D613-65, Test for Ignition Quality of Diesel Fuels by the Cetane Method [*Annual Book of ASTM Standards,* part 47 (1978)].

Diesel index is sometimes seen as a substitute for cetane number. It has the appeal that it is much easier to determine, but in consequence is a less accurate predictor of ignition quality.

8.2 FUEL SPECIFICATIONS

Confronted with a sample of fuel, one may ask, "which of the several fuels is this?" The answer may come from the *specification* it meets, a specification being a listing of limits (upper, lower, or both) on every property

130 FUELS AND LUBRICANTS

significant to the intended use of the fuel. Specifications are issued by several different organizations; in the United States the American Society for Testing and Materials is the prime issuer of specifications for a wide range of products, and is also the prime specifier of tests by which compliance with specifications is determined. The U. S. Navy also issues specifications that cover fuel properties (though none of the navy fuels are of interest to the users of low speed diesels), and some engine builders issue specifications for fuels to be used with their products.

The American Society for Testing and Materials (known as ASTM) publishes its specifications annually in a set of hardbound volumes [*Annual Book of ASTM Standards*]; specification D975 is the one for diesel fuels. It lists fuels designated 1-D, 2-D, and 4-D. These specifications for 1978 are reproduced here in Table 8.1.

Marine gas oil and *marine diesel oil* are commercial terms in worldwide use, and not names assigned by recognized specifications. The first of these, however, usually meets the ASTM 2-D specification and some samples of it may meet the U. S. Navy diesel fuel specification. The other is a more viscous distillate and varies more widely in quality, depending on the source of the crude and the practices of the refinery. The intermediate fuels likewise meet no recognized specification, save a generally understood viscosity level, such as intermediate 15 for viscosity of 1500 seconds Redwood at 100 F, or intermediate 35 for 3500 seconds. Residual oil is also not defined by a recognized specification, although many samples meet the requirements of No. 6 oil in the ASTM specification D396 for furnace oils, since the specification indeed intends that this oil be a residual. For this reason, residual oil is sometimes referred to as "No. 6."

8.3 FUELS FOR THE LOW SPEED DIESEL

An important economic objective in many engine applications is to burn the worst of the available fuels, since this fuel will also be the cheapest. Among marine diesels, the properties most likely to signify where a fuel lies on the worst-to-best scale are ignition quality (cetane number), viscosity, and contaminant content. And among the fuels available to marine engines, the *residual* oil is the worst, being characterized by low cetane number, high viscosity, and the highest concentration of contaminants. Among these properties, it is the contaminant concentration (and the nature of these contaminants) that is usually limiting in the low speed application, that is, presents the problem of greatest magnitude to the engine. However, residual fuels are generally usable by the low speed engine, and are indeed widely used, though only after certain upgrading processes that are feasible on shipboard.

TABLE 8.1 DIESEL FUEL SPECIFICATIONS OF THE AMERICAN SOCIETY FOR TESTING AND MATERIALS[a]

Grade of Diesel Fuel Oil	Flash Point, (°C), Min	Cloud Point [°C (°F)] Max	Water and sediment (vol %), Max	Carbon Residue on 10% Residuum (%), Max	Ash (weight %), Max	Distillation Temperatures [°C (°F)], 90% Point Min	Distillation Temperatures [°C (°F)], 90% Point Max	Viscosity Kinematic (cSt[g] at 40°C) Min	Viscosity Kinematic (cSt[g] at 40°C) Max	Viscosity Saybolt (SUS at 100°F) Min	Viscosity Saybolt (SUS at 100°F) Max	Sulfur (weight %), Max	Copper Strip Corrosion Max	Cetane Number,[f] Min
No. 1-D A volatile distillate fuel oil for engines in service requiring frequent speed and load changes	38 or legal (100)	[b]	0.05	0.15	0.01	...	288 (550)	1.3	2.4	...	34.4	0.50 or legal	No. 3	40[f]
No. 2-D A distillate fuel oil of lower volatility for engines in industrial and heavy mobile service	52 or legal (125)	[b]	0.05	0.35	0.01	282[c] (540)	238 (640)	1.9	4.1	32.6	40.1	0.50 or legal	No. 3	40[f]
No. 4-D A fuel oil for low and medium speed engines	55 or legal (130)	[b]	0.50	...	0.10	...	...	5.5	24.0	45.0	125.0	2.0	...	30[f]

Source: Reprinted with permission from the Annual Book of ASTM Standards, D975. (© American Society for Testing and Materials, 1916 Race Street, Philadelphia, Pa. 19103.)

[a]To meet special operating conditions, modifications of individual limiting requirements may be agreed upon between purchaser, seller, and manufacturer.
[b]It is unrealistic to specify low-temperature properties that will ensure satisfactory operation on a broad basis. Satisfactory operation should be achieved in most cases if the cloud point (or wax appearance point) is specified at 6°C above the tenth percentile minimum ambient temperature for the area in which the fuel will be used. The tenth percentile minimum ambient temperatures for the United States are shown in Appendix 2 of the source. This guidance is of a general nature: some equipment designs, use flow improver additives, fuel properties, and/or operations may allow higher or require lower cloud point fuels. Appropriate low-temperature operability properties should be agreed on between the fuel supplier and purchaser for the intended use and expected ambient temperatures.
[c]When cloud point less than −12°C (10°F) is specified, the minimum viscosity shall be 1.7 cSt (or mm²/sec) and the 90% point shall be waived.
[d]In countries outside the United States, other sulfur limits may apply.
[e]Where cetane number by Method D 613 is not available, ASTM Method D 976, Calculated Cetane Index of Distillate Fuel may be used as an approximation. Where there is disagreement, Method D 613 shall be the referee method.
[f]Low-atmosphere temperatures as well as engine operation at high altitudes may require use of fuels with higher cetane ratings.
[g]1 cSt = 1 mm²/sec.
[h]The values stated in SI units are to be regarded as the standard. The values in U.S. customary units are for information only.

132　FUELS AND LUBRICANTS

The term "generally usable" implies something less than "universally usable." Recall that residual oil is not a substance defined by recognized specification; its properties vary with source of the crude and with the practices at the refinery of its origin, but whatever they are, they are. A particular batch may *not* be usable by a low speed engine, meaning perhaps that it will cause fouling and corrosion-wear at an unacceptable rate. Now, the magnitude of "unacceptable rate" is subject to judgment, and this judgment differs among shipowners, and may differ between a shipowner and the builder of his engine. This situation is illustrated by Table 8.2, taken from a MAN publication [Beer (1978)]. Note that an upper limit of properties is stated, with recommended limits somewhat lower in several properties.

The 3500 sec viscosity recommended in Table 8.2 is typical of the low speed engine builders, and it is a value that precludes most undiluted residual oils. On the other hand, the 6000 sec viscosity limit (and the corresponding limiting values of the other properties as well) encompasses most residuals. A user who follows the recommendation must therefore buy mostly intermediate 35, a fuel consisting of residual diluted with enough distillate to bring it to 3500 sec Redwood viscosity at standard temperature. His engine therefore *does* burn residual fuel, but not in its "pure" form.

The term "heavy oil" is commonly used to describe the fuels used by low speed engines. The term is sometimes misunderstood to be synonymous with "residual," but properly means "residual oil, possibly diluted with sufficient distillate to suit the fuel property limits (usually viscosity) established by the owner." Semantically, that's about it; economically, the consequences of diluting the residual to 3500 secs Redwood is trivial, since the price of the intermediate fuel is not significantly different from that of the residual. An engine burning intermediate 35 is usually assumed to have essentially the same fuel costs as if burning undiluted residual. However, prices are continually changing; hence the validity of observations such as this may change also.

The heavy oils are universally distinguished by their viscosity, though both the first paragraph of this section and Table 8.2 indicate the significance of other properties. The principal reason for this is that viscosity is the most easily measured property, being measured by the time required for a standard quantity of oil to drain through a standard hole. A second reason is that the other properties are fairly well correlated by viscosity. For example, if the viscosity at standard temperature is less than 3500 sec, the cetane number can safely be assumed to be above the 25 listed in the table. The same can be said, though surely with greater uncertainty, about the contaminants. The residual oils have the highest sulfur fraction, for example, and reducing their viscosity by dilution with

TABLE 8.2 QUALITY REQUIREMENTS FOR HEAVY FUEL OIL AS PUBLISHED BY MAN

	Four-Stroke Diesel Engines						Two-Stroke Diesel Engines			
Engine Model	AS 25/30 32/36 GV 30/45 ATL		40/45		52/55A 65/65		KSZ A		KSZ A + B	
	Recommended	Spec. limit	Recommended	Spec. limit	Recommended	Spec. limit	Recommended	Spec. limit	Recommended	Spec. limit
Density at 15° C (g/cm^3)	0.920	0.940	0.970	0.980	0.980		0.980	0.990		
Viscosity at 100° F (sec Redw. 1)	200	1000	1500	3500	3500		3500	6000		
Sulfur content (wt %)	2.5	3.0	2.5	4.0	4.0		3.0	4.0		
Conradson carbon residue (wt %)	5.0	8.0	8.0	10.0	10.0		10.0	15.0		
Vanadium content (ppm)		<100		<200			—	—		
Sodium content (ppm)							—	—		
Water content (vol %)	<0.2	<0.5	<0.2	<1.0			<0.2	1.0		
Cetane number	>40	>40	>40	>40			>25	>25		
Flash point in closed cup (°C)	>65	>65	>65	>65			>65	>65		

Source: Beer (1978).

a distillate reduces this fraction also. But the residuals differ in sulfur content and in viscosity independently, so that producing a specified viscosity by dilution is no guarantee of a certain sulfur fraction. And the contaminants, such as water, that result from handling bear no relation to viscosity. Nonetheless, viscosity is the one index that most closely identifies a fuel, and it is often used in this way.

We have noted that the contaminant concentration is usually the factor limiting the use of a fuel. Some contaminants, such as water (if present in high concentration, at least) and gritty solids, can make a fuel unusable; on the other hand, such things are readily removable by filtering or centrifuging. Sulfur does not affect the immediate operation of the engine, but shows its deleterious presence in the long-term condition of the engine. The first product of sulfur is sulfur dioxide (SO_2), a gas which mostly escapes to the atmosphere with the other exhaust gas constituents. But one of these other constituents is water vapor, and at any point where the gas is cooled below its dew point, the vapor condenses to liquid with some of the gaseous products in solution. Other products of fuel contaminants—vanadium pentoxide (V_2O_5) is usually named as the chief culprit—act as catalysts for the reactions that lead from SO_2 to sulfur trioxide (SO_3), and then to sulfuric acid (H_2SO_4) when liquid water is present. This acid is corrosive to the metal surfaces, and does its dirty work wherever temperatures are low enough for the dew point to be reached.

The exhaust gas dew point will obviously be reached at some place downstream in the exhaust line, perhaps outside the ship, but otherwise perhaps within the exhaust line (for example, in the waste heat boiler) when the engine is running at light loads, or when the exhaust components have not warmed to operating temperature following a start-up. Some of the gas also contaminates the lubricant coating on the cylinder liner walls, and thereby being in intimate contact with this cooled surface, produces the acid end product. In consequence, a cylinder lubricant not designed to neutralize acids becomes corrosive to the surface it is supposed to protect; the result is an accelerated wear rate.

A general preventive measure for the corrosive events is to avoid the low temperatures that allow water condensation. This is in part a design responsibility—avoidance of over-cooled areas—and in part is an operational responsibility. Of course, the ship must maneuver at low powers, and the engine must start and stop, so that low temperatures cannot be completely avoided, especially at the lower end of the cylinder liners. A possible remedy is the use of a low-sulfur distillate fuel for all low-load running, and many ships have been provided with the requisite complication of fuel piping to allow switching of the main engine to the distillate fuel carried for normal use by high speed generator engines.

Whatever design and operational measures are taken against the formation of acid, its presence seems inevitable when high sulfur fuel is burned. The remaining weapon against its consequences within the engine is a neutralizing (that is, alkaline) cylinder lubricant. This is to receive further discussion in Section 8.5.

The other contaminants are notorious mainly for the deposits they produce within the engine. The deposits interfere with exhaust flows and with fuel spray patterns, clog piston ring grooves, and additionally may be corrosive to the surfaces they cover.

One class of deposit is that formed by unburned carbon. Now, carbon can scarcely be called a contaminant in a fuel which is principally carbon in composition, but these deposits do originate mostly in an identifiable part of the fuel, a part that can be removed aboard ship before the fuel is burned; hence their inclusion here among the contaminants is appropriate. This identifiable part, this effective contaminant, is the "gunk" part, the asphaltene compounds which in other contexts are used to pave roads. When the fuel is heated sufficiently to reduce its viscosity for injection, these compounds dissolve into the bulk of the fuel, and so pass readily into the cylinder, but their individual ignition lags are so great that they are nearly incombustible. One consequence is injector "trumpets," carbon deposits so-called because they resemble the bell of a trumpet, formed around the injector tips. They can grow sufficiently large to interfere with the injector spray pattern, thus exacerbating incomplete combustion. Carbon deposits similarly may appear within the exhaust ports, turbocharger turbine, and piston ring grooves. The common remedy is removal of the asphaltene fraction before use, although it apparently can be burned successfully if homogenized within the fuel, or if certain fuel additives are used. These matters are discussed further in the next section.

The other common class of deposit is that originating in the metallic contaminants, principally vanadium and sodium. (Vanadium is a natural contaminant; sodium originates principally from seawater contamination.) Their oxides, combined in complex fashion, and perhaps with participation from sulfur compounds, form an "ash" which is liquid above a temperature of about 500 C. When impinging upon metal surfaces above this in temperature, the ash adheres, and subsequently remains as a solid deposit when the surface cools. This deposit appears on the hottest cylinder parts, namely, the exhaust valve and its seat. (Builders of valveless engines sometimes point to their freedom from this hazard as an advantage.) Piston crowns (top surfaces) may also suffer. As a preventive measure, the engine designer provides cooling of the hot parts, and treatment of the fuel with additives may be beneficial. The latter point is discussed in the next section.

8.4 FUEL TREATMENT

Residual oil, whether "pure" or blended into an intermediate fuel, requires treatment before use by the engine. Treatment always includes heating to lower the viscosity, and mechanical removal (filtering or centrifuging, for example) of contaminants. It may also include contaminant removal by washing, and neutralization of deleterious behavior by additives.

Injection viscosity requirements depend on the particular engine in question, but typically are in the neighborhood of 100 sec Redwood, requiring a temperature in the neighborhood of 120 C. This temperature is readily attained in a heater supplied with low-pressure steam, usually the product of the engine's waste heat boiler.

The most common mechanical cleaning method uses centrifuging in either one or two stages to remove water and solid contaminants. If two stages (two centrifuges in series) are used, the first is the water remover (the "purifier") while the second removes remaining solids (the "clarifier"). Oils of very high viscosity usually contain the asphaltene "gunk" constituent mentioned in the preceding section, and this, too, is removed by centrifuging.

Washing consists of mixing hot fresh water into the oil at about 5 to 10 percent water to oil, plus subsequent removal of the water by a centrifuge. The centrifugal separation is usually enhanced by the inclusion of a de-emulsifying agent in the wash water. The purpose of the washing is to remove water-soluble contaminants; presumably during the brief intimate contact between oil and water any substance that is more soluble in water than in oil will be preferentially absorbed by the former, and so be removed by the centrifugal separation. Any contaminant that has originated in seawater (for example, sodium in the form of sodium ion) is likely to be removed in this way, but prominent contaminants such as sulfur and vanadium are not removed.

A thorough discussion of centrifugal purification of marine fuels and lubricating oils can be found in a paper by Trowbridge [Trowbridge (1960)] listed in the references for this chapter. Chapter 10 here has further information in the form of an diagram of a purification system.

Filtration is an alternative to centrifuging, though it is a later development (1970s) and not so widely used at the date of this book. Usually a filtration installation combines filters in series, with the first stage collecting the bulk of the contaminants in filters that can be cleaned by back flushing, and the following stage or stages collecting the fine particles in disposable cartridges. The last stage typically can remove water and solid particles of 5 microns in size. Some manufacturers claim removal down to 2 micron size.

A second alternative to centrifuging is coarse filtration followed by homogenization of the fuel. The homogenizer is essentially a grinder that reduces solid and liquid particles alike to very small dimensions and disperses them homogeneously throughout the fuel. The asphaltene constituents are among those broken up and dispersed, and the proponents of homogenization claim that they subsequently burn without causing the usual carbon deposits. Advantages of the process are that it allows burning of these parts of the fuel, and that it produces no residues requiring disposal. However, at the date of this book homogenization is still in the development stage.

In addition to the several methods of cleaning fuel, a wide variety of "additives" can be mixed with it in order to neutralize the behavior of contaminants that cannot be removed. For example, addition of small amounts of magnesium sulfate is sometimes used to reduce the ill effects of vanadium. It acts to raise the melting point of the vanadium-caused deposits, making them less likely to adhere to engine surfaces. But this is one example out of many possible additive applications; additives are sold by numerous sources with claimed amelioration of practically any recognized fuel defect. Judging from the many references to them in trade literature, their degree of success and the shipowners' opinions of their worth vary quite widely.

One safe generalization is that effective use of additives requires that they be added in the correct amount, and that they be uniformly distributed throughout the fuel. This means that adding/mixing equipment, most often a mixing tank with a metering pump (a small reciprocating pump that injects a known amount per stroke), must be part of the fuel system.

8.5 LUBRICANTS

The low speed diesel is unique in that it uses two distinct lubricating oils. One, the oil used by bearings (crankshaft bearings, crosshead bearings, for instance) is a bearing lubricant much like that used for a wide variety of applications; it is known as the "crankcase oil" or "system oil." The other, the oil used to lubricate the cylinder walls, is specific to this engine because it must be designed to interact with the product of sulfur combustion; it is appropriately known as "cylinder oil."

The inside of an engine cylinder is a rather severe environment for a lubricating oil, but the oil is continuously replaced as it is consumed. Feed rates are usually in the neighborhood of 1.0 g/kWh. The rate is under control of the operator, and therefore depends on the operator's conflicting desires to minimize the expense of lube oil consumption and to reduce cylinder liner wear rate by copious lubrication. The oil is usually a mineral oil (that is, obtained directly from constituents of refined petroleum rather than being a manufactured "synthetic") of SAE 50 viscosity index. Its most distinguishing property—established by additives—is its "total base number" (TBN), a property that measures its ability to neutralize acids. This number is the number of milligrams/gram content of base compounds equivalent to potassium hydroxide. A TBN of 70 is typical. However, the correct value of TBN is roughly a function of the sulfur content of the fuel; as this is written, lubricants of TBN up to 100 are available for use with fuels having especially high sulfur content.

The crankcase or system oil is usually a mineral oil also, but of lower viscosity, typically SAE 20 to 40, and of lower TBN, typically 15. A modest neutralizing ability, expressed by the TBN, is necessary because of the possibility of leakage into the crankcase of contaminated lubricant from the cylinders. The crankcase oil is also burdened by possible water contamination from a leaking oil cooler, and by high temperature if it is used for piston cooling. These stresses are combated by additives and by continuous purification. As examples of additives, oxidant inhibitors increase its resistance to oxidation by high temperatures, and "dispersants" keep solid contaminants from coagulating to form sludge deposits in the crankcase.

The system oil is continuously cleaned by centrifuging to remove water and solid contaminants. Water washing is sometimes used as part of the cleaning process, especially with the purpose of removing acids. However, washing is incompatible with acid neutralizing additives in the oil since they are likely to be washed out also.

8.6 GASEOUS FUEL

Ships carrying liquefied natural gas (LNG) form a small but significant part of the world commercial fleet. They are unique in that they continuously lose part of their cargo, this part being the vapor which forms—the "boil-off"—as heat seeps into the cargo through the surrounding insulation. Although this gas can be captured and reliquefied, or can simply be vented to the atmosphere, the most economical course is usually to divert it to the propulsion plant to supplement normal fuel. This alternative can be used with steam, gas turbine, and diesel plants.

The use of this gas in marine diesel engines was first put into practice by Sulzer Brothers [Steiger and Smit (1975), Sulzer Brothers (1968)] and the discussion here refers particularly to their engines. Figure 8.2 is a cross section of the cylinder of a Sulzer engine equipped for gas burning. Its unique feature is a gas admission valve; it is shown open to admit gas

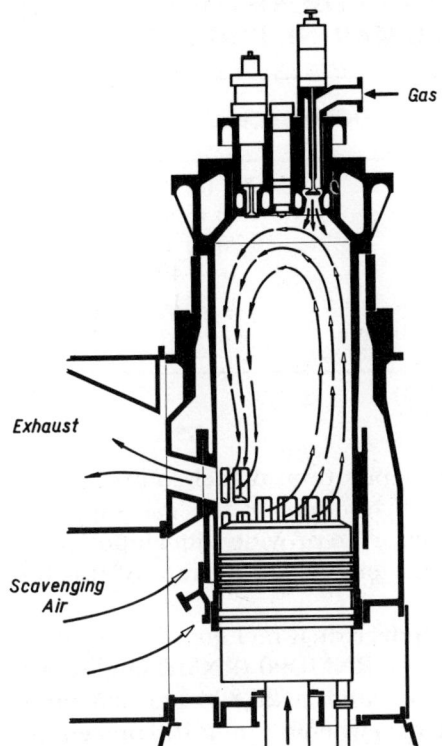

Figure 8.2 Cross section of Sulzer dual-fuel engine, showing gas injection during scavenging phase of piston cycle.

during the scavenging period. Note that the gas is directed against the incoming air, this to promote thorough mixing. As the piston proceeds into its compression stroke, it is a fuel-air mixture that is compressed, not air alone as in ordinary operation. This mixture is ignited by the injection of a pilot spray of the engine's normal oil fuel. The pilot fuel must be at least 5 weight percent of the total.

Whenever fuel is compressed with its combustion air (as in the ubiquitous spark-ignition engine), detonation or "knock" may occur from compression ignition, and this phenomenon limits the amount of gas that can be allowed in the cylinder and hence limits engine rating while burning gas. The tendency to detonate is a function of fuel composition, so that the derating depends on the gas composition; outside certain limits of composition the gas may not be usable. For example, limits published in 1973 [*The Motor Ship* (1973)] for Sulzer engines are given in Table 8.3.

The proportion of oil to gas fuel used simultaneously can vary from 100 percent oil down to the minimum ratio of 5 percent oil as established by the need for the oil pilot fuel. The ratio at any time depends on the

TABLE 8.3 LIMITS OF GAS COMPOSITION FOR SULZER DUAL-FUEL MARINE ENGINES

Gas	Limit (vol %)
Methane	60 to 100
Ethane	0 to 5
Propane	0 to 1
Nitrogen	0 to 40
CO_2	0 to 10
Higher hydrocarbons not allowed	

Source: THE MOTOR SHIP (1973).

amount of gas available from cargo evaporation. A fuel control system developed by Sulzer determines the magnitude of the gas supply by sensing its pressure, then adjusts oil supply to provide a fuel input necessary to meet the output demand of the engine. A diagram of this fuel control is shown in Figure 9.15.

The first ship to be powered by a Sulzer dual-fuel engine was completed in 1973 [*The Motor Ship* (1973)]. A 7RNMD90 (RNMD engine type of 7 cylinders, 900 mm bore) is used, rated at 20,300 bhp maximum continuous (2900 per cylinder) at 122 rpm when burning 100 percent oil, and 14,000 bhp (2000 per cylinder) at 122 rpm at 30 percent oil, 70 percent gas, the minimum oil/gas ratio allowed in this particular installation. The output at 122 rpm as a function of the gas-oil fuel ratio is shown in Figure 8.3.

The engine just cited is equipped with three auxiliary scavenging blowers, electrically powered by a shaft-driven generator, and operating in parallel with the piston underside scavenging pumps. These blowers replace the auxiliary low-load blower normally installed on a RNMD engine. The added air supply raises the mean effective pressure that can be attained without gas detonation.

8.7 REFERENCES

1978 *Annual Book of ASTM Standards* (1978), Part 23, American Society for Testing and Materials

Beer, P (1978), "Some Points Important when Comparing Diesel Engine Plants with Steam Turbine Plants for Ship Propulsion," presented to Institute of Marine Engineers, New York, April 26, 1978.

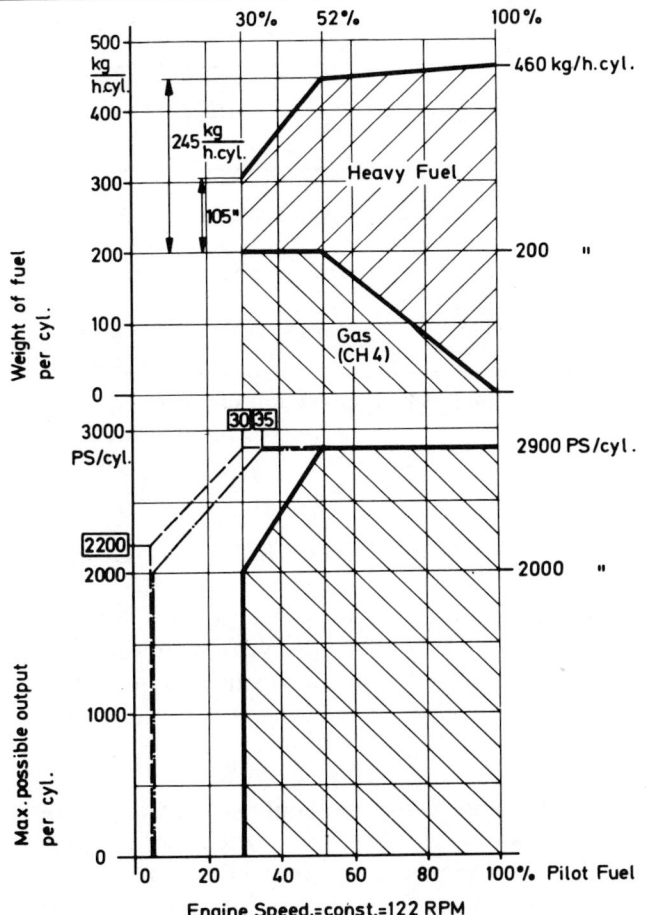

Figure 8.3 Output of Sulzer RNMD90 dual-fuel engine as a function of the oil–gas ratio [Steiger and Smit (1975)].

The Motor Ship journal (1973), "First Sulzer RNMD Type Dual-Fuel Diesel Engine Enters Service in LNG Ship VENATOR," Vol 54, 641, pages 423-428, December 1973.

Steiger, H A and Smit, J A (1975), "Development and Practical Application of the Large-Bore Direct-Drive Dual-Fuel Engine as Propulsion Unit for LNG Carrier," American Society of Mechanical Engineers paper 75-DGP-1.

Sulzer Brothers (1968), *Some New Developments in Sulzer Diesel Engines,* unpublished notes.

Trowbridge, M E O (1960), "Centrifugal Purificaiton of Oils for Marine Service," *Transactions,* Institute of Marine Engineers, Vol 72, pages 1-32.

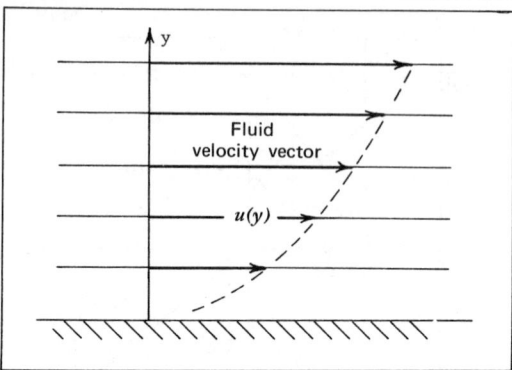

Figure 8.4 Explanation of viscosity.

8.8 APPENDIX - SOME THINGS ABOUT VISCOSITY

The shear force per unit area (τ), between layers of fluid in a viscous flow field (Figure 8.4) is proportional to the velocity gradient, du(y)/dy. The constant of proportionality (μ) in the resulting equation,

$$\tau = \mu \, \frac{du(y)}{dy} \tag{8.1}$$

is the *viscosity* of the fluid. Dimensions of viscosity are therefore force × time/area. The common scientific unit is the *poise*, or for more manageable numerical magnitudes, the *centipoise* (10^{-3}N s/m², or 1 mPa s).

Viscosity divided by density of the fluid in question is called *kinematic viscosity*. The common scientific unit is the *stoke*, or the *centistoke*, 0.01 stoke.

In practice, viscosity is measure by an *efflux viscosimeter,* a device in which a standard amount of the fluid being tested is allowed to drain through a standard hole. The time required is a function of kinematic viscosity and so is used to express the quantity directly in the time dimension, usually in seconds. There are several different viscosimeters in general use. The most common in liquid fuel and lubricating oil technology are the Redwood No. 1 and the Saybolt Universal (SSU = Saybolt Universal Seconds). The Redwood and SSU units appear in the specifications that have been quoted in this chapter, and in Figure 8.1.

Conversion equations from Redwood and Saybolt units (designated t) to centistokes are

$$\text{Redwood: } C_s = 0.26t - \frac{172}{t} \tag{8.2}$$

$$\text{Saybolt: } C_s = 0.22t - \frac{180}{t} \tag{8.3}$$

8.9 NOTATION FOR CHAPTER 8

ASTM	American Society for Testing and Materials
bhp	brake horsepower
cSt	centistokes
g/hph	grams per horsepower hour
g/kWh	grams per kilowatt hour
HMN	heptamethylnonane
kg/h-cyl	kilograms per hour per cylinder
MAN	Maschinenfabrik Augsburg Nurnberg
mm	millimeters
ppm	parts per million
PS/cyl	horsepower per cylinder
rpm, RPM	revolutions per minute
SAE	Society of Automotive Engineers
SSU	saybolt seconds universal
TBN	total base number
t	time, seconds, in viscosity formula
u	fluid velocity
y	dimension, width of flow passage
τ	shear force per unit area
μ	viscosity

Chapter Nine

CONTROL AND MONITORING

The fundamental functions of control are to maintain engine output—its torque (or power) and its speed—at a desired level, and to change it to another level when that is ordered by its operators. Control is effected by sensing the differences between the desired speed (or "set" speed) and its actual value, followed by the appropriate adjustments to the fuel injected per cycle. Note that "speed" is used in the sentence just preceding as synonymous with "output;" this can be done because the torque component of output is established by the characteristics of the load—fix engine speed and the torque (or power) output must be that of the propeller or other load at that speed. The fuel adjustment can be done manually, or by some "automatic" device, such as the engine governor. In most cases, a governor and other control hardware lie between human commands and the fuel adjustment. Starting, stopping, changing speed, and changing direction (that is, reversing) are obviously necessary also, and are accomplished by a combination of operator commands and powered control actions.

The ship speed and direction must be controlled also, and this is usually synonymous with control of the engine (insofar as direction implies only forward or reverse, of course). However, the characteristics of the ship (its propeller as well as its hull) influence engine behavior, and so must be considered in design of engine controls. For example, the torque and speed imposed on the engine by the propeller during a reversing transient strongly influence the starting of the engine in the reverse direction. If the propeller is a controllable pitch unit, then the control of the ship speed is accomplished by control of shaft speed, or by control of pitch, or by a combination of both; control of the engine must then be integrated with that of the propeller.

The "monitoring" of engine conditions is closely related to control, since operator actions with respect to the engine are likely to be based on knowledge of these conditions. For example, a decision to overhaul a cylinder may be based on records of temperature measured by thermocouples embedded within the cylinder liner. And the practicality of leaving an engine unattended, and controlled therefore by nonhuman agencies, is established by adequate display and alarm of vital engine parameter values in remote locations where humans are on duty.

The preceding points mentioned are discussed in this chapter as well as certain special topics that are of interest. For example, the control of fuel flow when both gas ("boil off" from a liquefied natural gas cargo) and oil are used as fuel is described. An appendix outlines the principles of governing, a matter that is at the heart of engine control.

9.1 ENGINE CONTROL

To control an engine is to obtain a desired level of its output, and thereafter to maintain that level, or to change it as the demands of the load on the engine may require. Although "output" is a combination of torque (or power) and rotational speed, the latter is the variable actually controlled directly, since engine and load characteristics always interact so that obtaining amd maintaining a speed guarantees that torque or power requirements of the load are met. Rotational speed is also easy to measure, and is closely related to the variable ultimately to be controlled, namely the speed of the ship. To control an engine is therefore to control its speed.*

The only variable that can reasonably be manipulated to control speed is the fuel flow, more specifically the amount of fuel injected for each engine power stroke. This amount of fuel in most fuel systems is proportional to the effective stroke length of the fuel pump plunger(s). The essence of engine control is thus a sensing of rotational speed, coupled with the appropriate setting of fuel pump output (sometimes designated by "fuel rack position").

---

*An exception to this statement is the need to limit power, irrespective of speed, in order to avoid overload. Such a need can occur, for example, with a controllable pitch propeller.

A common method of setting the fuel pump is by means of a *governor*, a device that senses engine speed (see discussion in the appendix to this chapter), and reacts to adjust the pumps to maintain speed equal to its setting. The speed setting is obtained from a remote signal originating at a control station.

A simpler method is to set the fuel pumps directly from a human command, with no subsequent adjustments in response to speed variations. This alternative is called "fuel control," in contrast to "speed control" by the governor. Its use does not violate my statement that to control an engine is to control speed, since the human operator selects the fuel setting in anticipation that it will result in a desired speed. With a fixed-pitch propeller "fuel control" is acceptable because of the stable interaction of propeller and engine characteristics: an inadvertent increase in shaft speed produces a sharp increase in torque demand (torque proportional to square of speed according to the "propeller law"), thereby driving speed downward toward its original value. And the opposite occurs in reaction to a speed decrease. However, protection against overspeeding (as from operator error, for instance) must be provided, hence a governor that will override the manual fuel setting is always installed. This so-called "simple governor" acts to limit fuel if the maximum engine speed is reached.

For every existence there must be a beginning, of course, so that control of the starting process is an essential. The large engines of which we speak in this book are started by injection of compressed air into several (or all) cylinders; at least three must be equipped for air injection so that the engine will never be stuck on center when a start is attempted. The essential control functions are to open and close the cylinder starting air valves with the correct timing, and to admit air to the starting air main during the starting cycle. Timing of the cylinder valves is usually accomplished by a distributor valve, actuated by a camshaft, that admits air to the pilot valve that then opens the cylinder valve during the proper interval.

Low speed engines are reversible engines, hence a means of effecting reversals must be included in the control equipment. Direction of rotation is determined by the direction in which the engine is started, but each direction requires the same sequence of admission, injection, and exhaust, a condition that will *not* be met by simply reversing camshaft rotation. The phase of the camshaft with respect to the crankshaft must be changed, or a new set of cams can be brought into engagement by moving the camshaft axially a short distance.*

---

*If the fuel system is controlled by a microprocessor, that unit determines injector timing, and no camshaft is involved.

Reversing from a running state also implies stopping and restarting in the opposite direction: hence fuel and starting air must be manipulated in proper sequence with the cam-shifting operation: fuel to the cylinders is interrupted, the camshaft is shifted, starting air is applied (although the engine may still be rotating in the original direction at this point), and fuel flow is reinstated.

Figure 9.1 is a diagram of a control scheme embodying the notions expressed above. It is perhaps the simplest possible scheme, this said because it uses direct (that is, without the intervention of a governor) setting of fuel injection stroke, and does not include multiple control stations (no control from the bridge of the ship). The control console—presumably placed close to the engine—is in the lower right corner of the figure. Note the handwheel for direct fuel control, and the lever that actuates the starting, stopping, and reversing sequences.

The fuel pumps are represented in the figure by one set of pump-actuating linkages (item 9). Observe the cam (on the dashed line representing the camshaft) that actuates the linkage, the direct setting of linkage position (hence pump stroke length) from the console handwheel, and the cut-off servomotor (item 10) that can lift the linkage to stop injection. The servomotor is activated by either the overspeed governor (item 11) or the injection cut-off distributor (item 16) that acts when the lever is in the "stop" position or during the shifting of the camshaft

Five pneumatic pilot valves are shown within the console, each positioned by a cam on a camshaft rotated by the control lever. For instance, the ahead start position sends a pneumatic signal to position the starting air distribution servomotor (item 6) to the ahead position. This valve in turn positions the distributor for ahead rotation. The air starting valve (item 5) is likewise positioned via its servomotor (item 3). The same signal puts the camshaft into ahead position via components 8 and 14.

Figure 9.2 is a photograph of an engine control stand mounted on its engine. Components and their functions are similar to those in the console in Figure 9.1.

9.2 COMPUTER CONTROL

At the time this book was written, the microprocessor had not been applied to control of low speed engines in marine service. However, concurrent developments by engine builders indicated that it might indeed be placed into practice after 1980. For example, the MAN "electronically controlled injection" described in Chapter 3 (see Figure 3.24) is based on an electronic governor (built around a microprocessor) that computes a beginning and duration of injection that will optimize engine

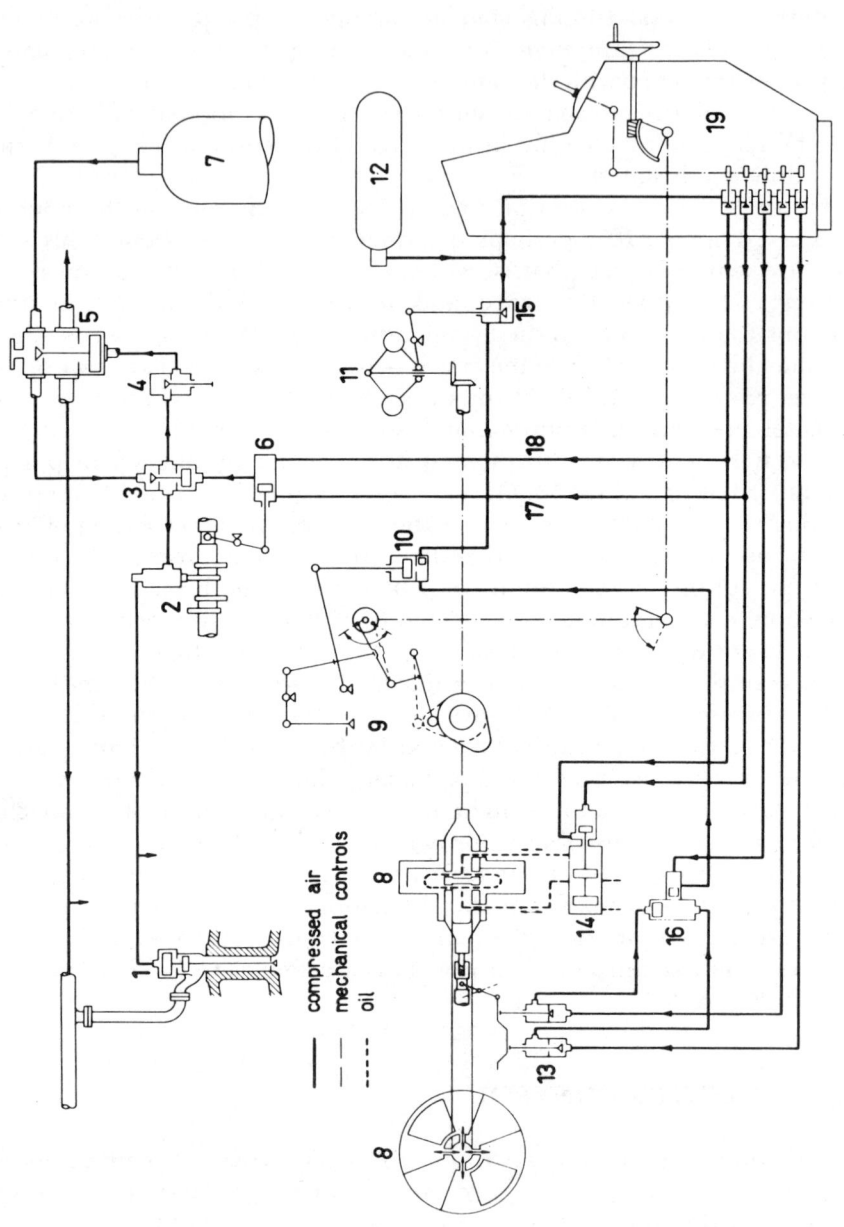

ENGINE CONTROL FROM REMOTE LOCATIONS 149

performance under current operating conditions, and sends a signal to the electro-hydraulic fuel pilot valve at each cylinder.

Engine speed is the most important of the conditions sensed by the processor; the speed signal comes from a magnetic pickup on the end of the crankshaft. Ambient air temperature is another of the conditions sensed. By a change in injection timing, the increase in fuel rate that may be caused by such factors can be kept to a minimum.

Starting and reversing functions are also computer controlled in the MAN system, since the starting valves can receive their actuating signals from the microprocessor. No camshaft need be reversed, since cams are not used to actuate the fuel injection valves or the starting valves.

The major advantage claimed for computer control is generally better performance under varying operating conditions, this because the computer can consider and act on any operational factor that can be sensed. A consequent benefit is a lower minimum speed. Because the injection period can be adjusted to suit speed, an engine can operate down to about one-sixth its rated speed, compared to approximately one-fourth with conventional control.

9.3 ENGINE CONTROL FROM REMOTE LOCATIONS

Engine control is traditionally managed from a control station (such as the console in Figure 9.1) located close by the engine. At the time this was written, however, this tradition had been pretty well displaced by a new tradition of control from remote locations, these locations usually being enclosed rooms overlooking (via a large window) the engine room, and the ship's bridge. Although the engine must still be manipulated in the ways outlined in the preceding section, additional complexity of component and system necessarily lies between engine and the human

---

Figure 9.1 Engine control schematic diagram [from Drake (1951)].
1 Starting valves on cylinder heads
2 Starting air distributor
3 Main starting valve servomotor
4 Air shutoff valve on turning gear
5 Main starting valve
6 Starting air distributor servomotor
7 Air reservoir (30 kg/cm^2)
8 Reversing gear
9 Injection pumps
10 Injection pump cutoff servomotor
11 Overspeed governor
12 Air reservoir (10 kg/cm^2)
13 Valves for fuel cutoff during reversal
14 Reversing gear oil cutoff valve
15 Overspeed governor servomotor
16 Injection cutoff distributor
17 "Ahead" starting
18 "Astern" starting
19 Pneumatic valves in maneuvering console

Figure 9.2 Engine local control stand (Sulzer photograph).

decision maker. For example, a deck officer commanding a reversal from the bridge cannot be expected to have the understanding of engine behavior possessed by his engineering colleagues. The engine must be "automated" to a degree sufficient to replace the senses and knowledge of an engineer who would otherwise be standing close by the engine.

The remote control scheme therefore is additionally complex because simple mechanical linkages are impractical over large distances, because additional logic* must be included to replace direct human interaction with the engine, and also because more than one remote location is likely to be required. With regard to the last, the engineers must have a control station even though control can be carried out from the bridge, and local control is necessary to protect against possible control system failure.

Among the several engine builders the remote arrangements of control differ with respect to use of pneumatic, hydraulic, and electrical components, with respect to the type of engine governor, and in other ways. And, of course, changes with time should be expected ("progress"). All of the variations must provide the necessary remote actuation capability, the built-in logic that the remotely located human cannot provide, and safety features. As an example of the last, an overspeed trip feature, set to stop the engine at 20 percent overspeed, independent of governor action, is always included because of regulatory body stipulations.

The Sulzer SBC-7 bridge control system is a typical example of engine control, and is pictured in several figures here. A block-diagram view is shown by Figure 9.3, and Figures 9.4 and 9.5 are schematic representations of the system. Figure 9.4 (Sulzer Dwg 107.068.437) does not include the bridge components, but contains the principal elements of control; the bridge features are added elements that are shown by 9.5. Figure 9.4 is the one to look at for the details. Now, you surely will not be able to divine the functions of all components and lines on either 9.4 or 9.5 (I couldn't!), but considerable understanding can be had by a study of them with the aid of the following description of 9.4, a description quoted from Sulzer literature [Sulzer Brothers, undated]:

Setting the Required Direction

The engine direction is selected manually by the reversing lever (8.42) of the pneumatic maneuvering unit. Moving this lever, say, to the ahead po-

---

*"Logic" refers to built-in relationships among steps in an expected sequence of events. For example, the cut-off of fuel during the reversing sequence mentioned in the discussion of Figure 9.1 occurs because of the built-in interaction between camshaft position and the cutoff valves. "Interlock" also describes such interactions, particularly when protection from operator error is involved.

152 CONTROL AND MONITORING

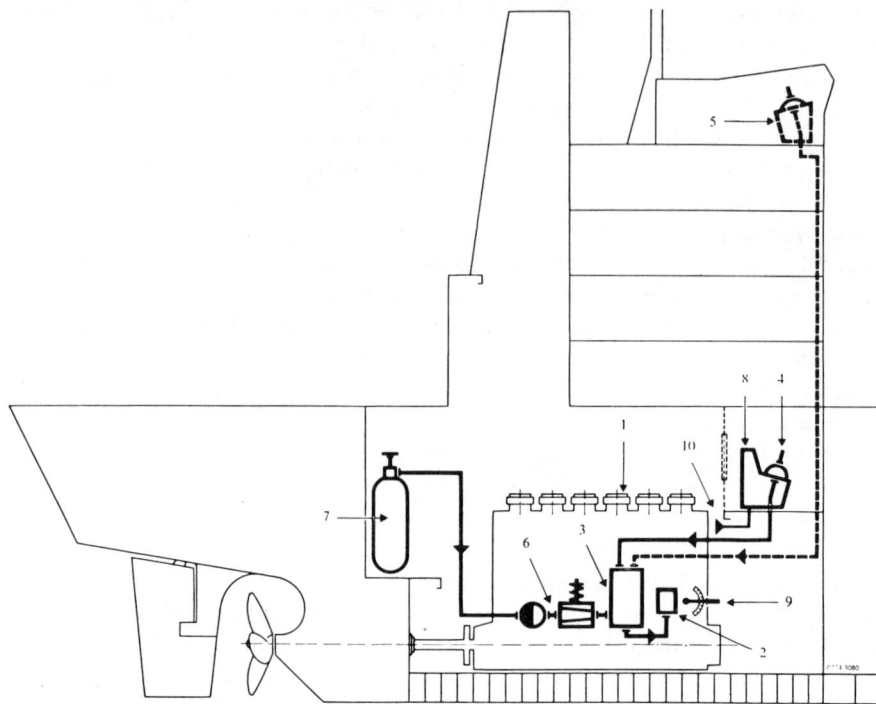

Figure 9.3 Schematic of Sulzer SBC 7 bridge control system (Sulzer drawing).
1 Main engine
2 Governor
3 Control box
4 Main engine maneuvering unit
5 Automatic engine control from bridge
6 Air filtering and pressure reducing unit
7 Starting air vessel
8 Central control desk
9 Emergency lever for independent mechanical control
10 Electrical supplies

sition actuates the corresponding pilot air valve (8.53) and a 7×10^5 Pa air signal is sent to the three position cylinder (8.02) and to the direction safeguard valve (8.45). The three position cylinder (8.02) then moves the reversing control valve (4.02) to the "ahead" position. Oil at 5×10^5 Pa from the main lubricating oil system flows through the valve (4.02) to the servomotor (4.01), turning the camshaft to the ahead position. When the servomotor reaches its end position the oil pressure builds up in the system and operates the hydraulic/pneumatic interlock valves (8.47) and (8.-48). As the interlock valve (8.45) has been previously activated by the "ahead" oil pressure from the reversing control valve (4.02), a signal can flow to valve (8.46). Starting air from start button (8.15) is thereby released.

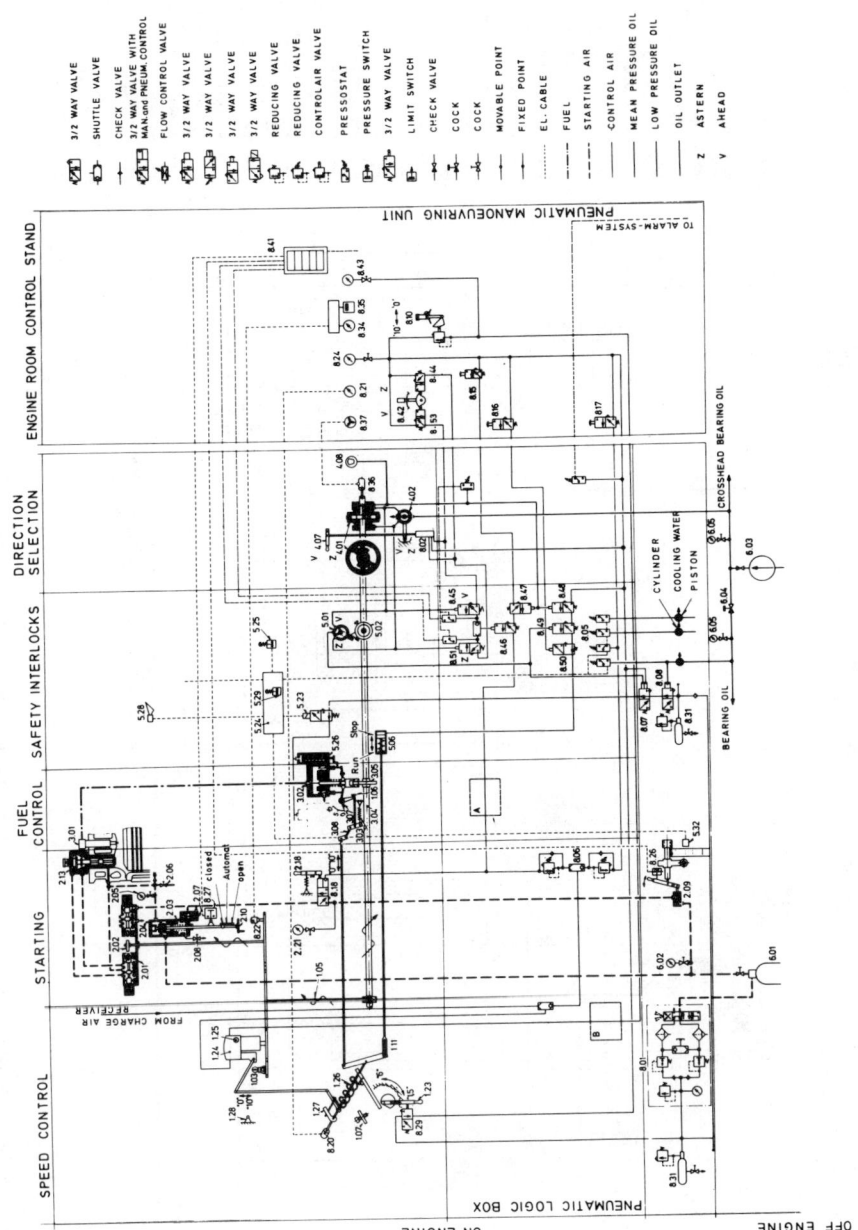

Figure 9.4 Schematic diagram of Sulzer SBC-7 engine control system (Sulzer drawing).

153

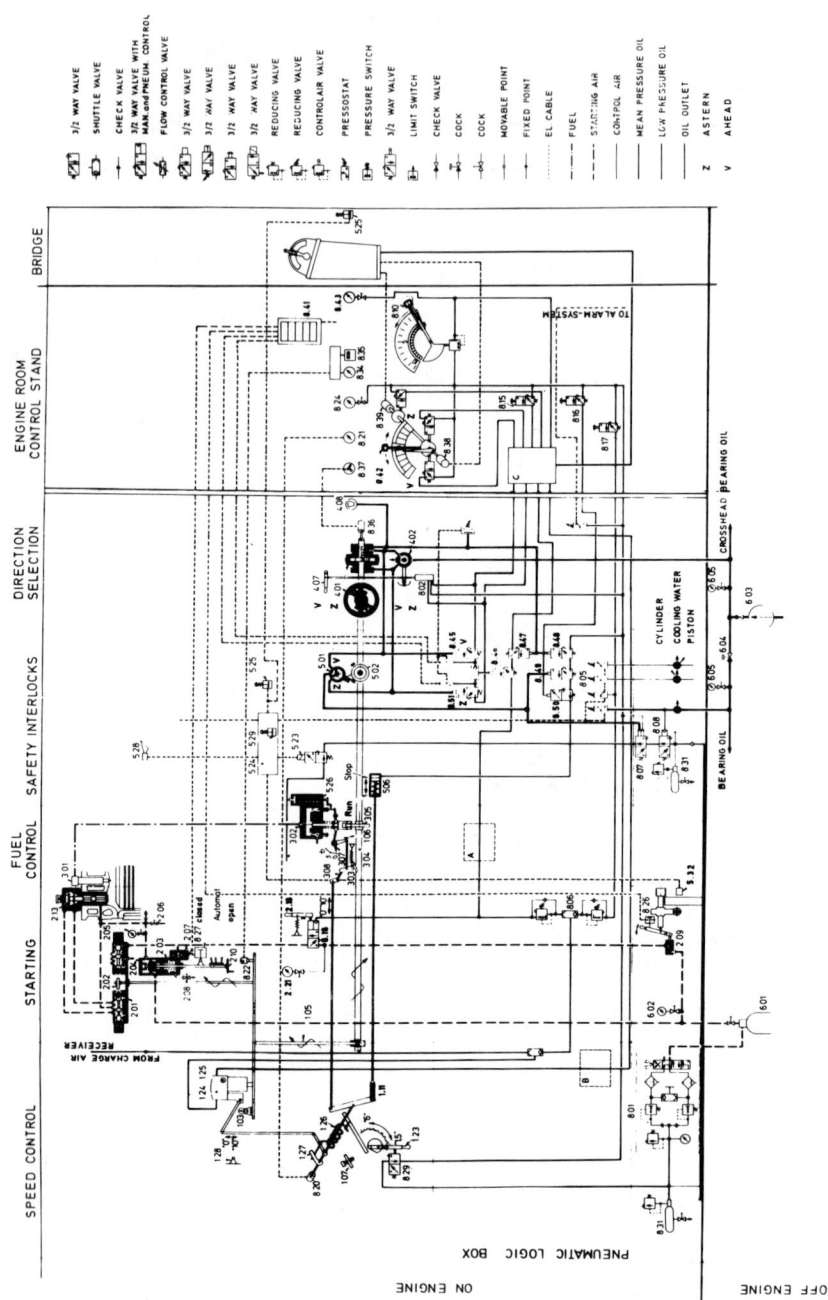

Figure 9.5 Schematic diagram of Sulzer SBC-7 engine control system, including remote bridge control (Sulzer drawing).

ENGINE CONTROL FROM REMOTE LOCATIONS

Fuel Setting

The speed setting lever (8.10) on the pneumatic maneuvering unit is moved to the starting position, approximately position 3.5 of a full range of 10 (in the RND engine fuel is admitted simultaneously with the starting air), thereby setting the governor input (1.24) to the same position.

Starting the Engine

By pushing the starting button (8.15), starting control valve (8.18) is opened allowing air at a maximum pressu of 30×10^5 Pa to flow from the air reservoir (6.01) through the interlock valve (2.09) on the turning gear to the main automatic starting air stop valve (2.03) and to the starting air distributor (2.01). The main automatic starting air stop valve (2.03) is opened and supplies air to the starting air valves (2.13) situated on each cylinder.

The starting air distributor cam (2.02) is driven from the engine camshaft and depending on its angular position actuates the appropriate starting pilot valve which opens the corresponding starting air valve (2.13) in the cylinder cover, thereby admitting air to that cylinder. As the engine turns the sliding coupling for the reversing safety interlock device (5.02) on the camshaft moves the reversing interlock (5.01) into the ahead position. If the engine is turning in the correct direction, oil will flow from (4.02) through valve (5.01) to the relay air valve (8.49), which then relays an air signal to the shutdown servomotor (5.06) which consequently releases the fuel rack and allows the governor to control the fuel. It will be seen that the engine cannot receive fuel before (5.01) and (5.02) are moved in this direction selected by (4.02). If the direction is wrong, that is, if in our example the propeller is still turning the engine in the astern direction, then the shutdown servomotor (5.06) is not released and the fuel rack is held in the stop position.

Further Points on Maneuvering

The reversing servomotor (4.01) is moved to the selected direction at any engine speed. This means that reversing can take place even with the engine turning at its full speed. Due to the particular design of the starting air valve (2.13) in the cylinder cover, braking air is applied as soon as the reversing action has finished. The reversing time required is approximately 5 to 6 seconds.

In an emergency reversing maneuver, the reversing lever (8.42) is put into the "astern" position when the engine is still running in the "ahead" direction. The fuel is immediately cut off by the shut-down servomotor

156 CONTROL AND MONITORING

(5.06) since the astern direction has been selected by the reversing control valve (4.02), but the engine is still turning ahead, leaving the direction of the reversing safety interlock system, (5.01) and (5.02) still in the ahead position. The reversing servomotor (4.01) will move to the astern position and the interlock valve (8.47) will free the starting button (8.15), so that when it is operated the engine will receive the "astern" starting air, thereby braking it to stop. As mentioned above, in an emergency case this braking air can be applied as soon as the engine is reversed into the new direction.

When the engine is brought to a stop it will immediately start to move in the astern direction. The reversing safet interlock system (5.01) and (5.02) will move in the "astern" direction and shut-down servomotor (5.06) is actuated so that the engine will once more receive fuel. When the starting button (8.15) is released, starting control valve (8.18) is closed, thus venting the starting control signal to the main automatic starting air stop valve (2.03) and to the starting air distributor (2.01).

Safety Shut-Off Features

The integrated control is provided with an electronic overspeed shut-off unit (5.24) fed from contactless pick-up (5.32). In case of overspeed the solenoid valve (5.23) is energized, which vents the spring-loaded cylinders (5.26) thus lifting the fuel pump suction valves and stopping the engine.

An automatic pressure cut-out system (8.05) is included. If the pressures of the piston or cylinder cooling water, bearing oil or control air fall below certain pre-set values, then an electrical signal is fed to the overspeed and safety cut-out unit (5.24) which stop the engine in the same manner as overspeeding.*

The additional features required for bridge control are shown in Figure 9.5 (Sulzer Dwg 107.068.438), though not in detail. You will readily observe the bridge control stand added at the right of the figure; closer inspection shows squares labelled A, B, and C that contain additional components within the "pneumatic logic box" of Figure 9.4.

Square A contains valves required for an automatic starting sequence.

Square B contains speed control programming components that stand between the bridge order and the order actually received by the engine governor. For example, a slow increase in engine speed may be wanted. Such an increase can be programmed to occur following a quick movement of the control handle, this to save the necessity for a human hand

---

*Regulatory bodies such as the American Bureau of Shipping require shutdown upon overspeeding, and upon low lube oil pressure [American Bureau of Shipping (1978)].

to remain on the handle, slowly nudging it from one position to another over a period of, say, a half-hour.

Square C components effect transfer of control between engine room and bridge, and perform certain additional safety interlocking functions.

Figures 9.6 and 9.7 are Sulzer drawings of the bridge control console. All communication between this unit and the engine room is pneumatic, though some electrical control occurs locally. The control handle functions in a manner similar to a conventional engine-order telegraph in that it has discrete positions "full," "half," "slow," and "dead slow." To each of these corresponds a definite maneuvering-range engine speed. The full ahead setting can be increased automatically to its at-sea value by the "square B" program mentioned previously, with the programmed time being as long as 40 minutes. Slowing to maneuvering speed is programmed likewise. Each speed is also adjustable by "fine tuning" buttons so that ship speed can be set at any value needed for travel in congested waters.

A deck officer takes control by turning a selector switch to the proper position. However, the order must be acknowledged from the engine control console by movement of its telegraph (item 8.42 in Figure 9.5) to the "bridge control" position. An engineering officer can retake control at any time by moving the telegraph out of this position.

Figure 9.8 is a photograph of the pneumatic "logic box" that contains most of the components essential to remote control of the engine. Note that it is outlined in both Figures 9.4 and 9.5.

Figure 9.9 shows schematically the Woodward PGA governor used on Sulzer RND engines equipped with the control system outlined here. If you have read the appendix to this chapter, and have examined Figure 9.19 there, many aspects of this diagram should be comprehensible without explanation. With this expectation, we note here only those aspects particularly related to Sulzer's application (some not discernible in the diagram). They follow, in outline format:

> Speed is set by a pneumatic signal from remote locations, with the speed set proportional to air pressure. A pressure of about 16 percent of full-speed pressure sets engine idling speed; zero pressure shuts off fuel.
>
> A manually activated device mounted on the governor allows speed to be raised to 120 percent of its rated value for the purpose of testing the engine's overspeed trip.
>
> The maximum allowed power piston travel is a function of both engine speed and inlet manifold air pressure. The speed function cuts

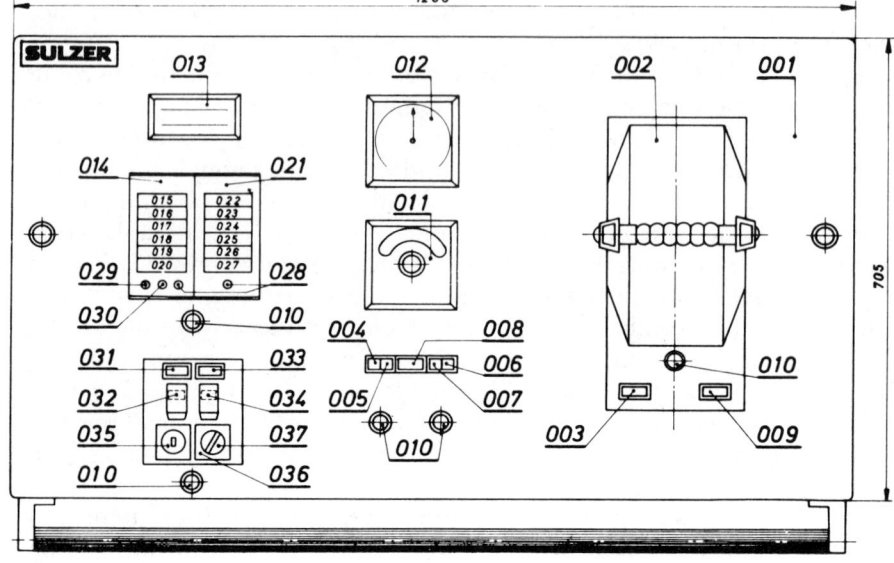

Figure 9.6 Sulzer bridge control console (top view) (Sulzer drawing).

001 Bridge console
002 Sulzer telegraph
003 Standby — Button
004 Fine setting "up" — Button
005 Fine setting "down" — Button
006 Automatic acceleration — Button
007 Automatic deceleration — Button
008 Automatic load program "off" — Button
009 Finished with engine — Button
010 Dimmers
011 Rpm preselection
012 Engine tachometer
013 Starting and control air gauge
014 Alarm panel
015 Engine overloaded — Alarm
016 Engine tripped — Alarm
017 Engine failed to start — Alarm
018 Starting air too low — Alarm
019 Control air too low — Alarm
020 Electrical power failure — Alarm
021 Indicating panel — Alarm
022 Bridge control — Lamp
023 Control room control — Lamp
024 Engine emergency control — Lamp
025 Transfer of controls — Lamp
026 Spare — Lamp
027 Automatic slow down — Lamp
028 Lamp test — Button
029 Stop horn — Button
030 Stop flash — Button
031 Emergency stop — Lamp
032 Emergency stop — Button
033 Emergency run — Lamp
034 Emergency run — Button
035 Main key switch
036 Selector switch (bridge conntrol/control room control)
037 Transfer of controls — Lamp

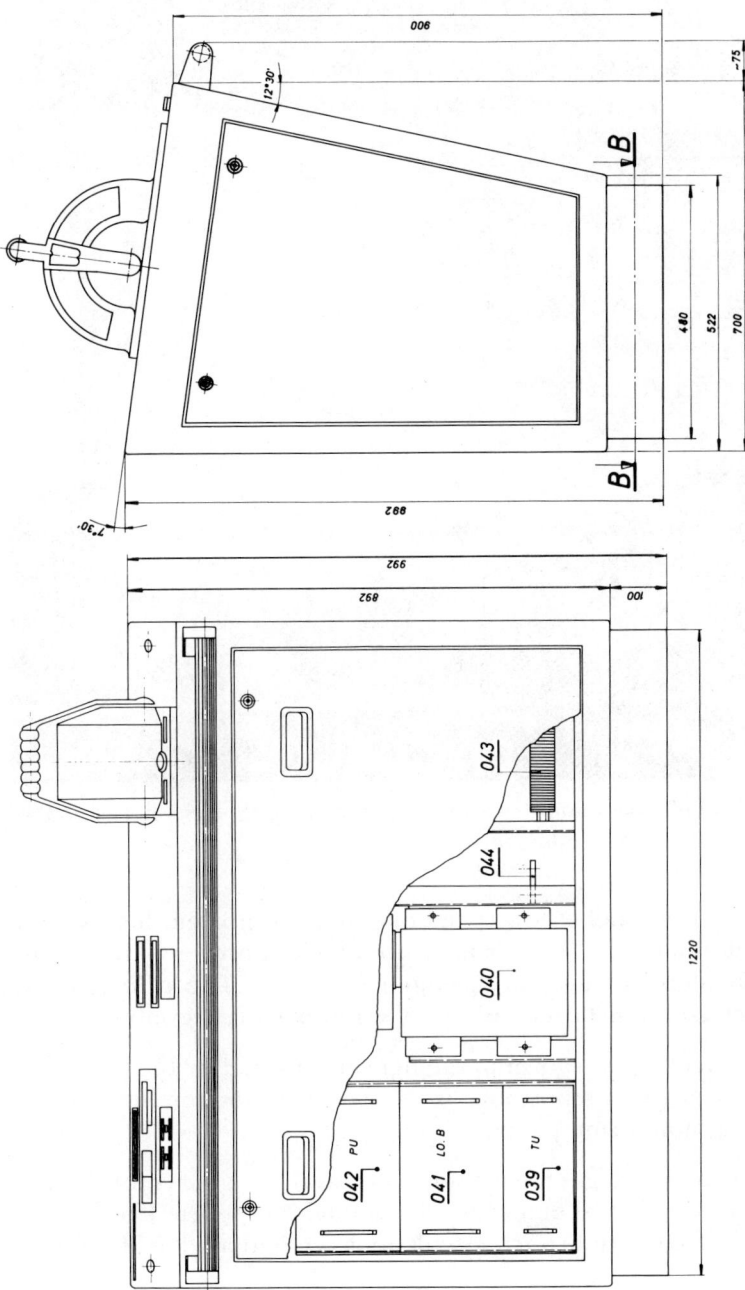

039 TELEGRAPH UNIT
040 LOAD PROGRAM UNIT
041 LOGIC BOX
042 ELECTR POWER UNIT
043 ELECTRICAL TERMINALS
044 PNEUM BULKHEAD CONNECTIONS

Figure 9.7 Sulzer bridge control console (elevation) (Sulzer drawing).

160 CONTROL AND MONITORING

Figure 9.8 Pneumatic logic box for engine control (Sulzer photograph).

back the allowed fuel at low speed operation to prevent high-torque, low speed, operation. The air pressure function prevents the air/fuel ratio from falling too low during engine transients. These functions act simultaneously; whichever calls for the lower limit prevails.

Microswitches (not shown in the figure) activated by the power piston tail rod signal its position to an overload monitor (indicator or alarm on a monitoring panel).

Speed-droop linkages are included so that the engine control can be combined with control of a controllable pitch propeller. The function of droop in controllable pitch installations is mentioned in Section 9.5.

Pre-loaded buffer springs reduce the response of the governor to

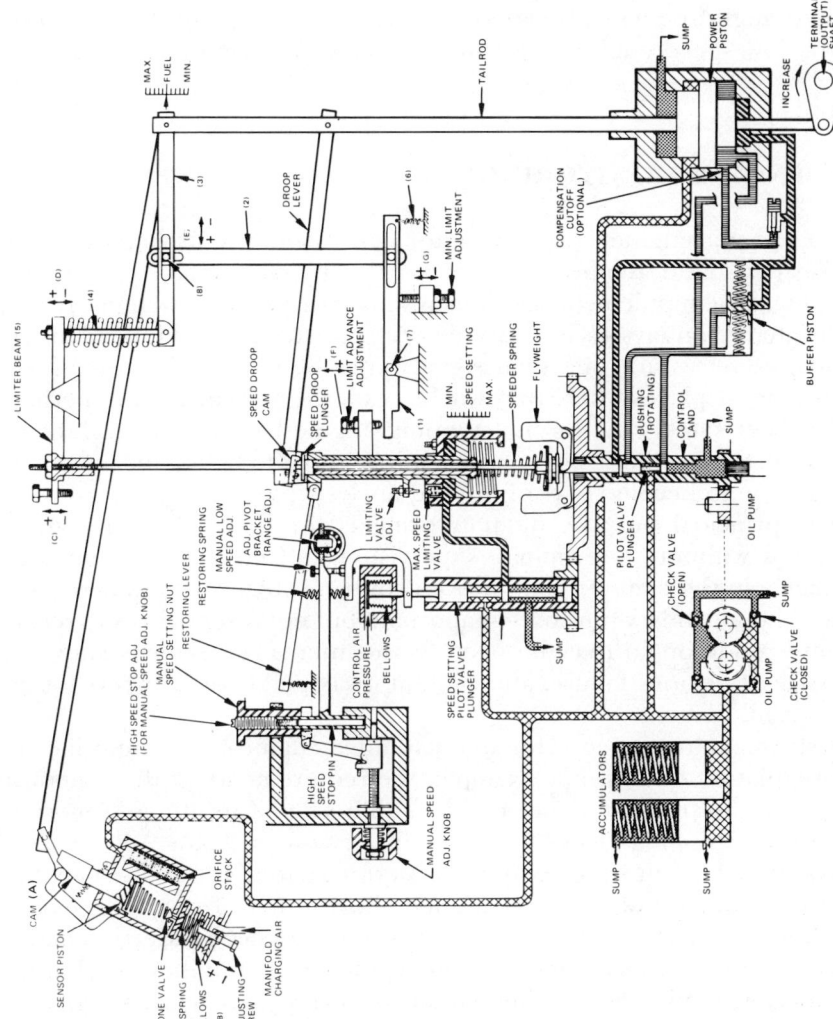

Figure 9.9 Schematic diagram of Woodward PGA governor used with Sulzer SBD-7 control system (Woodward Governor Company drawing).

162 CONTROL AND MONITORING

small high-frequency speed fluctuations, such as those that can occur in a low speed engine as a consequence of its firing impulses.

9.4 ENGINE MONITORING

An operating engineer can "monitor" an engine by its impact on his senses (its sound, temperatures, etc), aided by instruments (thermometers, etc) that supplement the human sense organs. If this engineer is to be located remotely—as is nearly always the case—then additional instruments are necessary, and their signals must be conveyed to the one or more remote places where they will be read, will activate alarms, perhaps will trigger safety shutdowns (stopping engine because of low lube oil pressure, for example), and perhaps will be recorded ("logged") for later reading or to become part of a trend analysis plot.

The principal purpose of monitoring is to allow the engine to be operated without close human supervision, indeed without human attendance in the engine room or even control room, for long periods of time. The secondary purpose—important, but not essential—is to record engine condition so that the need for maintenance becomes known in advance of obvious failure, and without disassembly of cylinders for examination.

Instrumentation needed for an unattended engine room is specified by the regulatory bodies. For example, the requirements of the American Bureau of Shipping are specified in Table 41.1 of its Rules [American Bureau of Shipping (1980)], and are listed here in Table 9.1.

Additional monitoring equipment is not required, but is often applied for the secondary purpose mentioned above—that of reporting engine conditions that demonstrate the present or future need for maintenance. This generally means detection and reporting of conditions within the cylinder, where lie the wearing parts that must be periodically replaced. A simple strategy (not requiring monitoring) is to make replacements on a fixed schedule based on expected life of these parts. A similar strategy is to make examinations (requiring disassembly) on a fixed schedule, with replacements made on the basis of the results. Both are imperfect, for they inevitably entail labor and/or use of parts that is unnecessary when the wear rate doesn't fit the fixed schedule. If, on the other hand, wear of a cylinder liner (for example) can be detected and recorded continuously, it will be replaced only when it has reached the maximum tolerable wear, and no time and labor will be expended on examinations before that point is reached.

TABLE 9.1 AMERICAN BUREAU OF SHIPPING REQUIREMENTS FOR MONITORING IN UNATTENDED ENGINE ROOMS

	Display	Alarm
Fuel to engine	Pressure	Failure
Fuel to engine	Viscosity	High
Fuel drain tank	Level	High
Fuel purifier	—	Malfunction
Lube oil to engine	Pressure	Low
Lube oil to turbocharger	Pressure	Low
Cylinder exhaust	Temperature	—
Engine exhaust	Temperature	High
Starting air	Pressure	Low
Coolant expansion tank	Level	Low
Cooling medium	Pressure	Low
Cooling medium (out)	Temperature	High
Oil mist detector	—	High density
Scavenge air box	—	Fire
Bearings (main and thrust)	Temperature	High

(Oil mist in the crankcase is an indication of being overheating.)

Source: Information in this table is extracted from Table 41.1 of the American Bureau of Shipping *Rules for Building and Classing Steel Vessels* (© 1980 American Bureau of Shipping). The *Rules* are updated annually; hence the table is subject to change.
Consult the latest edition to learn if any changes have been made in the classification requirements.

The notion of using monitoring of engine condition to predict a future maintenance action is pictured by Figure 9.10. The ordinate may be total wear of a component such as a cylinder liner; the points are the readings taken periodically (every week, every 100 hours running time, etc).

Wear—if it can be measured continuously—is an obvious choice, and is easily translated into a plot such as Figure 9.10. One method of wear measurement is to examine lube oil for metal content, this on the grounds that any metal present must be the consequence of wear, but the method suffers from uncertainties in translating readings into wear rates. A later method (developed in the 1970s) is based on direct sensing by a sensor that itself suffers the same wear as the component being monitored. Figure 9.11 shows a sensor that is inserted through a cylinder liner wall so that it is worn by the passing piston rings in the same manner as

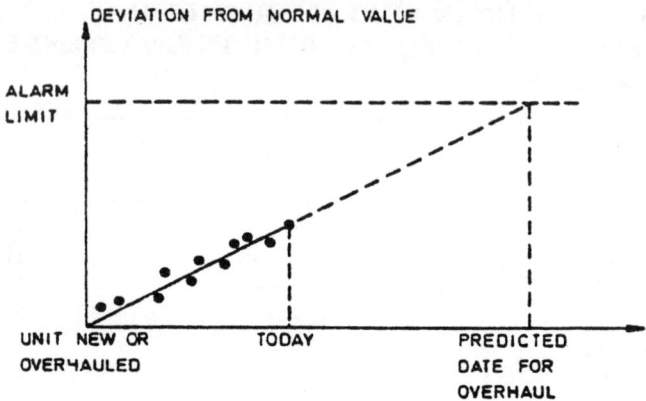

Figure 9.10 Predicting overhaul date by monitoring [from Langballe et al (1975), *Norwegian Maritime Research*, Vol 3, 3]

the wall itself. The "wear sensor element" of the figure is a thin-film resistor. As it is worn away, its resistance changes, an occurrence that easily translates into an electrical signal.

Temperatures also are significant to cylinder condition in several ways. For example, sharp transient jumps in temperature on the liner surface as a piston ring moves by are indications of the wear phenomenon known

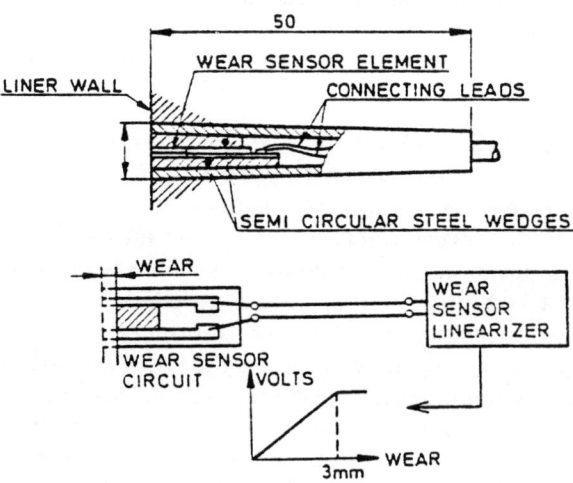

Figure 9.11 Details of a cylinder liner wear sensor [from Langballe et al (1975), *Norwegian Maritime Research*, Vol 3, 3].

as "scuffing." A thermocouple penetrating the surface (much like the wear sensor of Figure 9.11) can report the temperature jumps. Although a quantitative measure of wear is not possible from this, it alerts the operating staff to the condition, a condition that often can be cured immediately by increasing the flow rate of lubricant to the affected cylinder.

Thermocouples placed within holes that penetrate to within 5 to 10 mm of the inner surface of a cylinder liner report temperatures that are very nearly those of the cylinder wall surface, and so give indication of cylinder conditions of a less local nature than scuffing. If there is no other truly local distortion, then the temperature so measured is a good indicaton of "thermal load" (rate of heat flow per unit area) experienced by the cylinder boundaries. Abnormal temperatures may be caused by abnormalities in fuel injection or air supply; if the latter is also monitored (flow and temperature), a defect in fuel injection (distortion of spray pattern, say) may be deduced from temperature readings. Local overheating caused by such things as piston-ring blow-by may be apparent also if liner thermocouples are placed in the appropriate location (near the position of the top ring when the piston is at top center).

The thermal load on cylinder boundaries can also be monitored by thermocouples in the cylinder cover, and this location has the advantage that local effects such as piston-ring blow-by are not likely to distort the general picture. Figure 9.12 shows a thermocouple placed in a hole drilled into a cylinder cover.

Another important cylinder-status item is the condition of the piston rings. Blow-by (escape of cylinder gas past the rings) and ring scuffing are defects that temperature signals can detect, as just noted. Additional information can be had from sensors consisting of small induction coils. If these are inserted through the cylinder walls in a fashion similar to that of the wear sensors, a pulse is generated as each ring passes the sensor location, and condition of the ring can be deduced from the appearance of the pulse on an oscilloscope screen. If, for example, a pulse is significantly lower in height than the others, then the ring that it represents is probably sticking in its groove, or broken.

Monitoring of air flow by means of a pressure difference across some part of the air intake path, such as the turbocharger inlet diffuser, provides useful information about the air system (clogged cylinder inlet ports, for example), and can point to the cause of cylinder defects detected by the other sensors.

If the engine room is to be unattended, then the monitoring specified by the regulatory body must be done. In addition, monitoring of fuel pressure, injection valve timing, and cylinder pressure is sometimes applied.

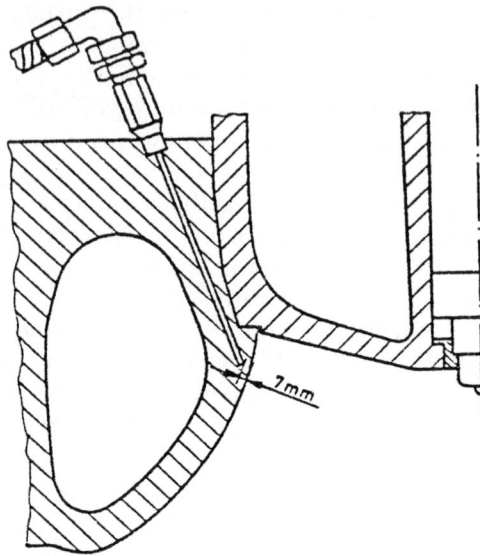

Figure 9.12 Thermocouple in engine cylinder cover [from Langballe et al (1975), *Norwegian Maritime Research*, Vol 3, 3].

Such are the items monitored, with our view being concentrated on the monitoring points themselves; the circuits that transmit and process the signals, and the display or recording hardware, we shall regard as beyond the scope of this text and leave to other sources of discussion. The major source used here is a paper by Langballe et al [Langballe et al (1975)] which gives an excellent summary of the monitoring techniques of the late 1970s.

This section closes with a description of a typical organization of the monitoring signals that are required by regulatory bodies (Table 9.1). The scheme described is that used by MAN in the 1960s and following, as reported by Scobel and Richter [Scobel and Richter (1968)]. Other manufacturers may do it differently, but the MAN scheme typifies what must be accomplished for an unattended engine, and so serves as a good model.

The safety monitors produce alarms to alert the distant human operators to the need for corrective action. There are three alarm groups: STOP, REDUCE SPEED, and ENGINEER, listed in descending order of urgency. The STOP alarm is tripped by (1) failing lube oil pressure, (2) deviation of exhaust gas temperature of one cylinder from the mean temperature of all cylinder exhausts, and (3) high bearing temperature. The alarm is given in the engine control room, on the bridge, and in the stateroom of the engineer on watch, though the last two locations can be disabled when the control room is manned. It is intended that manual shutdown be the response to this alarm, but in addition, the low lube oil pressure also produces an automatic shutdown unless a deck officer gives an override by pressing an emergency maneuver button on the bridge.

The REDUCE SPEED alarm is tripped by (1) high exhaust gas temperature, (2) high cooling water temperature, and (3) low cooling water pressure. The alarm is given in the same locations as the STOP alarm.

The ENGINEER alarm requires no emergency action, and so is not given on the bridge. However, it must be acknowledged, and if it is not acknowledged within 3 minutes, a general alarm is given on the bridge (oh, where are those lazy engineers!?). The items covered by this category of alarm are all of the regulatory ones not otherwise alarmed, and any others that the shipowner may have wanted to include.

9.5 CONTROL WITH A CONTROLLABLE PITCH PROPELLER

When the pitch of a propeller is fixed, there must be a unique relationship between power and engine speed; the power-speed operating point must lie along the propeller characteristic. If, on the other hand, the pitch can be changed, this characteristic can be shifted so that the operating point can lie anywhere on the power-speed plane, though bounded by the extremes of pitch available to the propeller, and the extremes of mean effffective pressure and speed of the engine. Within these bounds, independent manipulation of pitch and engine fuel position moves the operating point at the will of the operator.

But the will of the operator may inadvertently produce operating points that are less desirable than those of the fixed-pitch propeller (touched on previously in Section 5.6); hence a definite program of power-speed combinations is usually incorporated into the engine-propeller control system.

Figure 9.13 illustrates in simple fashion several possible ways of imposing such a definite program. Note that there is a single control handle which rotates two cams. As the handle is moved, these cams each act upon a pneumatic valve (typically) to generate a signal (pressure) to the corresponding controlled device. The relationship between the two signals—the "program"—is obviously determined by the shapes of these two cams. In the left sketch, a speed (N) signal is sent to the engine governor, which sets the fuel level (Q) to produce that speed. A pitch signal (P) is sent directly to the pitch servomotor. If maximum allowed fuel setting is reached, an overload signal (Qmax) reduces pitch.

Other schemes are possible, including the one shown by the right-hand

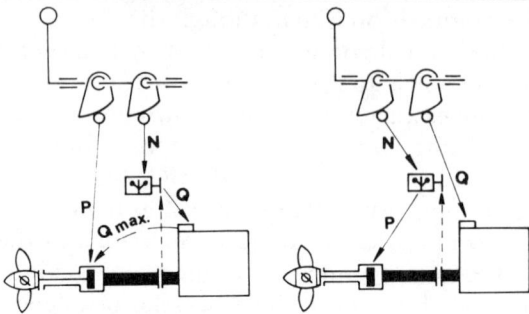

Figure 9.13 Engine–propeller control program expressed by cam shapes (two alternatives shown) [from Schanz (1967), Society of Naval Architects and Marine Engineers, by permission].

sketch. Perhaps more common than either one shown in Figure 9.13 is the use of a pitch governor and an engine governor (and other engine control components as discussed earlier). Each governor receives its speed-setting signal, and acts to maintain that speed, the engine governor via fuel adjustments, the pitch governor via pitch adjustments. In a scheme of this nature developed by Escher Wyss (Escher Wyss, Ravensburg, Germany) the speed setting to each unit is nominally the same, but the pitch governor is isochronous, while the engine governor has a significant droop. The resulting speed is the one at which the respective governor speed-load curves intersect, as shown in Figure 9.14. The intersections form a locus across the power-speed plane; in this way the power-speed program planned by the ship designer is obtained.

When a pitch governor and an engine governor are used, the former usually is adjusted to respond first to an upset in speed. Thus, for example, an increase in resistance due to bad weather produces an automatic pitch reduction to maintain shaft speed, so that the engine speed and power are unaffected. Recall that this feature of controllable pitch operation has been discussed in Chapter 5.

At very low powers, the programmed relationship between engine and propeller is usually replaced by direct pitch control. The engine remains at its minimum speed under control of its governor; speed signals generated by the bridge control handle are interpreted as demands for a particular pitch setting.

The interaction of engine and controllable pitch propeller is discussed well and thoroughly in a 1967 Society of Naval Architects and Marine Engineers paper [Schanz (1967)]. Our discussion here is condensed from that source, and it is recommended for careful reading.

The low speed engine is always a reversible engine, and the controlla-

FUEL CONTROL WITH LIQUEFIED NATURAL GAS FUEL 169

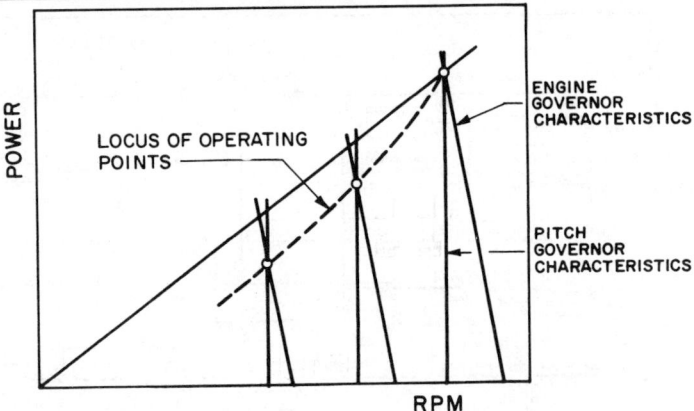

Figure 9.14 Engine–propeller control program developed by intersection of engine and propeller–governor characteristics [adapted from Schanz (1967)].

ble pitch propeller is a reversible propeller via its ability to reverse pitch. When the two are used together, the usual arrangement is to reverse by reversing the propeller. The engine reversing controls are "blanked off," but are available for emergency use at the engine manual control station.

9.6 FUEL CONTROL WITH LIQUEFIED NATURAL GAS FUEL

Engines that propel ships carrying liquefied natural gas are usually expected to burn the gas that evaporates from the cargo tanks ("boil off"), but this fuel does not replace the oil, it only supplements it to the extent that gas is available. The engine cylinders are therefore equipped with both the normal injector and a gas admission valve (recall Section 3.7 and Figure 8.5). Control of fuel flow to the engine must encompass simultaneous flow from two diverse sources.

Figure 9.15 illustrates a fuel control scheme used by Sulzer in such applications [Steiger and Smit (1975)]. The engine governor sets both the oil fuel pump stroke and the gas pump stroke via the linkages labelled (8) and (9). If, for example, there is no gas pressure, the spring (12) holds the control rack of the gas actuator pump (11) in its extreme position, making the bottom of (9) a fixed pivot, so that all governor motion acts solely on the oil injection pump (10). If, however, there is sufficient gas pressure, the fuel divider (14) pushes the fuel rack to the left, reducing oil fuel injection. Since the governor's command for fuel is unchanged, the position of the midpoint of link (9) is unchanged, and this link pivots about that midpoint to increase the stroke of the gas actuator pump.

170 CONTROL AND MONITORING

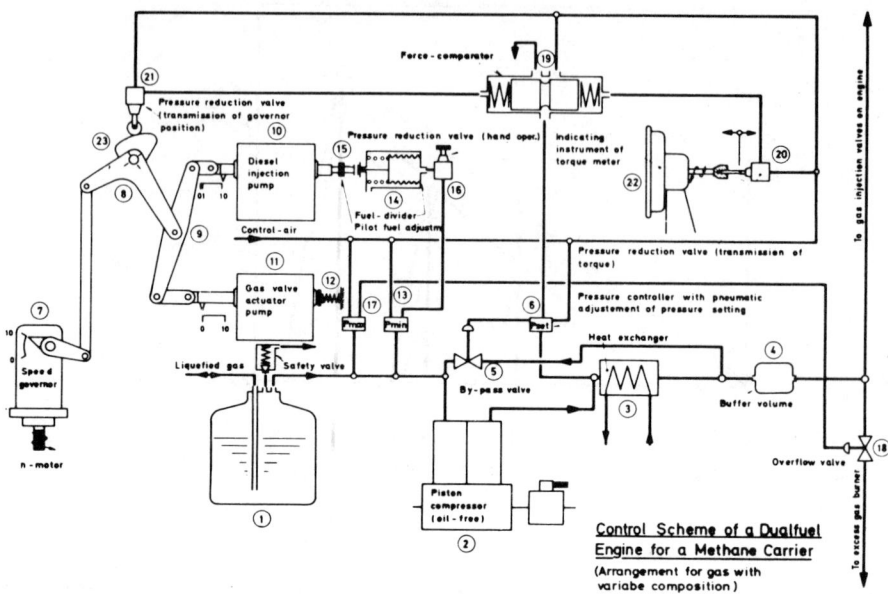

Figure 9.15 Schematic diagram of combined control of gaseous and liquid fuels [from Steiger and Smit (1975)].

Power thus remains constant, but the split between the two fuel sources is altered to suit the availability of gas. If for some reason the operators do not want to use all of the gas, adjustment of hand valve (16) changes the balance toward oil by changing the pressure that acts on the bellows (14). The consequent rising gas pressure opens the overflow valve (18) to dispose of the excess gas.

The system also contains features to compensate for varying heating value of the gas, a factor that can vary widely in some cases. The force comparator (19) compares governor demand via valve (21) to engine torque via valve (20). A deficiency in the latter causes the comparator to adjust pressure controller (6) upward, effectively increasing gas flow.

9.7 REFERENCES

American Bureau of Shipping (1980), *Rules for Building and Classing Steel Vessels*.

Drake, Forrest (1951), "The Governing of Internal Combustion Engines," *Proceedings,* American Society of Mechanical Engineers Oil and Gas Power Division.

Langballe, M, Tonning, L, and Wiborg, T (1975), "Condition Monitoring of Diesel Engines," *Norwegian Maritime Research,* Vol 3, 3.

APPENDIX TO CHAPTER 9 - GOVERNING   171

Schanz, F (1967), "The Controllable Pitch Propeller as an Integral Part of the Ship's Propulsion System, "*Transactions* Society of Naval Architects and Marine Engineers, Vol 75, pages 194 - 223.

Scobel, H H, and Richter, J H (1968), "A New Approach to Maintenance and Operation of Large-Bore, Two-Diesel Engines, and Experience in Operation of Periodically Unattended Engine Rooms," *Transactions*, Society of Naval Architects and Marine Engineers, Diamond Jubilee Meeting, pages 18-1 - 18-20.

Shipbuilding and Marine Engineering International, midsummer 1976 issue, page 375.

Steiger, H A, and Smit, J A (1975), "Development and Practical Application of the Large-Bore Direct-Drive Dual-Fuel Engine as Propulsion Unit for LNG Carriers," American Society of Mechanical Engineers, paper 75-DGP-1.

Sulzer Brothers (undated), "Integral Engine and Bridge Control System for Sulzer RND Engines."

Welbourn, D B (1963), *The Essentials of Control Theory for Mechanical Engineers*, American Elsevier Publishing Company.

Welbourn, D B (1963), "The Governing of Diesel Engines in Ships," *Shipbuilding and Shipping Record,* International Marine Design and Equipment Number, 1963.

9.8   APPENDIX TO CHAPTER 9 - GOVERNING

The *governor* is the traditional device for automatically maintaining constant (or nearly constant) the speed ordered by the operator of an engine. Its history long predates that of the diesel, since governors exhibiting some of the features of present-day units were used by early steam engines.

A typical governor consists of three basic elements: (1) a speed-sensing device, (2) an amplifier to increase the small forces generated by the speed sensor to usable magnitude, and (3) a servomotor to move the fuel rack position.

Engine speed can be sensed in several ways, classifiable as electrical, mechanical, or hydraulic. The traditional method is the mechanical flyball, and it is still the most commonly used method. See Figure 9.16. The whole assembly rotates at a speed proportional to engine speed; inspection of the figure shows that centrifugal force in the flyballs produces couples about the pivot points that are balanced by couples produced by the spring force. Position of the base of the spring is a function of the initial spring force and of the speed. For a given initial spring force, which is the operator's way of setting a desired speed, there is hence a unique position for every speed. The output force somewhat upsets this relationship, especially if the force is a mechanical friction. Nonetheless, this force is sometimes used in simple governors to set the fuel position

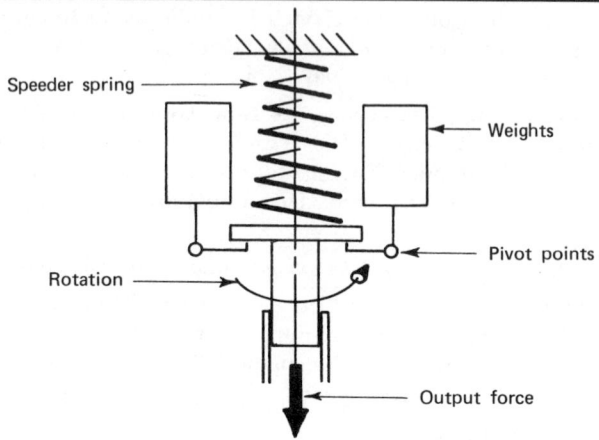

Figure 9.16 Elements of a flyball governor.

directly. In such cases, sensor, amplifier, and servomotor functions are combined into a sigle device. These governors are commonly called *mechanical governors.* Alternatively, the output force moves a hydraulic pilot valve (the amplifier) which in turn directs oil under pressure to an actuating cylinder (the servomotor), thus providing what is usually called a *hydraulic* governor. In a hydraulic governor, the amplification allows the flyball output force to be quite small compared to that needed by a mechanical governor.

Three qualities should be expected of a governor: (1) force or torque output adequate to position the fuel setting, (2) adequate speed of response, and (3) stability (the engine should not "hunt" about its set speed). These, especially the last two, involve engine characteristics as well as characteristics of the governor itself. Understanding of the problems involved must be based on knowledge of at least the basics of automatic control theory. All of this is beyond the scope of this text; hence I beg off from giving a dissertation on these aspects of governing. Readers with further interest might turn to a book by D B Welbourn [Welbourn(1963)], a control theory textbook that treats the governing of diesel engines.

On the other hand, a fair understanding of governor behavior can be had from brief discussions of the essential elements of these devices. Look at Figure 9.17, a figure showing these elements for three types of hydraulic governor. The first (upper left) shows the simplest device, one that produces an *integral* action. Observe that oil flows to the power cylinder as long as any error in speed exists to keep the pilot valve open. The movement of the power piston (that which moves the "throttle") is

APPENDIX TO CHAPTER 9 - GOVERNING   173

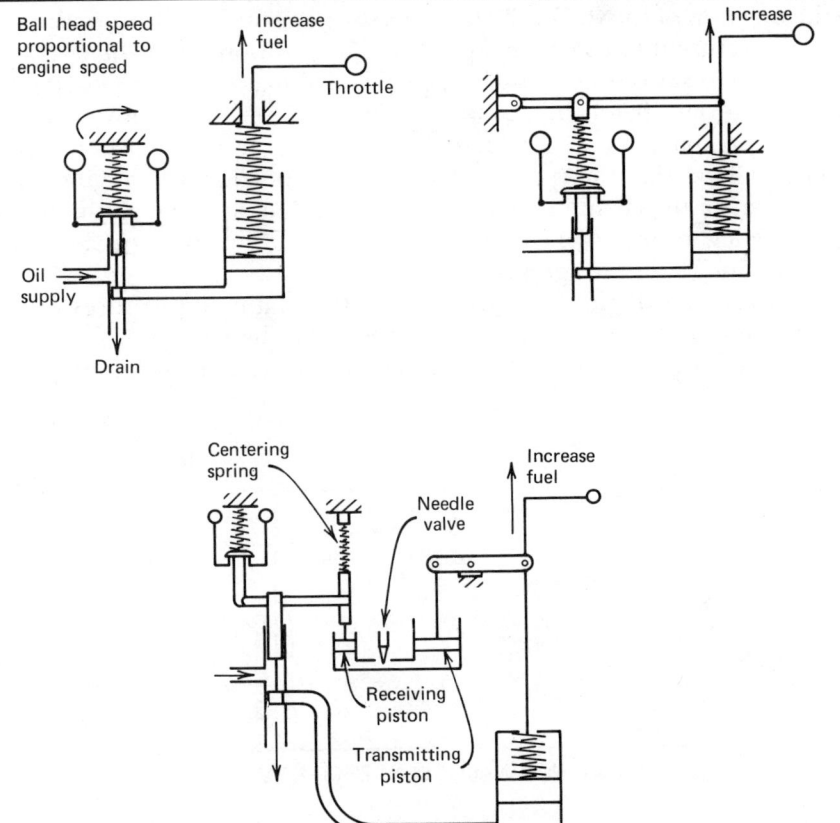

Figure 9.17 Elements of three types of flyball governor [from Drake (1951)].

thus the integral over time of speed error; hence the name for this governor's mode of action. Unfortunately, stable control is not likely here, because there is no feature to prevent engine speed from increasing or decreasing past the set speed before the flyballs stop correcting for the error.

The second sketch (upper right) shows a governor in which the instability of unmodified integral action is eliminated by a lever connecting governor output to the spring that balances flyball position (the "speeder spring"). As the power piston moves upward to increase speed, the lever moves the speeder spring in a direction that relieves its force, thus tending to close the pilot valve. Movement of the power piston is thereby slowed as the set engine speed is approached, and a slow but stable approach is the consequence. Another consequence is a resultant offset in the position of the top of the speeder spring. The action just described does not return it to its original position (the position set by the operator to obtain a desired speed); hence actual speed maintained by the governor may deviate from the set speed. The position of the power piston must be proportional to the load on the engine; from the action of the lever it follows that the speed error is also proportional to engine load. The governor thus exhibits *proportional* action. This action is also denoted by the inelegant but explicit term *droop*. The consequences of droop are explored in a later paragraph.

The third sketch shows a governor that contains the features of the second, plus additional elements that allow it to be *isochronous,* meaning that it maintains the set speed without droop as the load changes. It

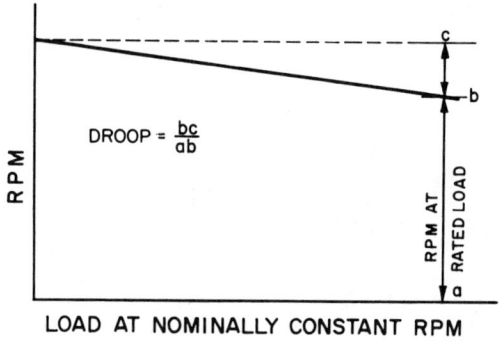

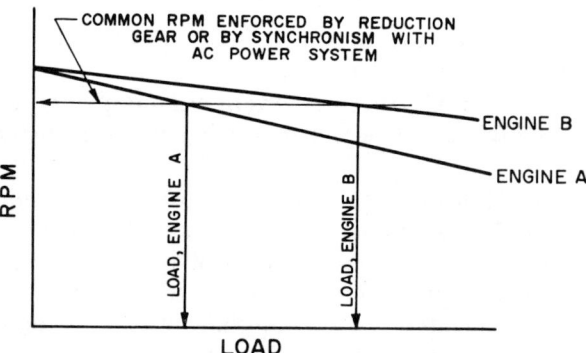

Figure 9.18 Governor droop and its effect on load sharing between two engines.

obtains stability via a transient droop, however, because the needle valve allows only a slow leakage of oil. Note, then, that the assembly of the receiving and transmitting cylinders, and their links to the power piston and flyballs, acts as a single rigid lever, except as modified by leakage past the needle valve. The governor thus responds to a rapid speed change (or change in speed setting) just as the one in the second sketch does. Over a longer period of time (on the order of seconds), the escape or inflow of oil allows the centering spring to slowly restore the original speed setting. This last is *reset* action; hence the governor is spoken of as having *proportional plus reset* action.

The meaning and utility of *droop* can be amplified by a look at figure 9.18. The formal definition of the term is given in the top sketch of the figure. The definition is based on the assumption—typically found in practice—that the speed-load curve expressing droop is linear. The second sketch indicates how droop is exploited to establish load sharing among engines driving the same load, as in a multi-engine propulsion shaft, or paralleled diesel generators. The sketch actually shows mismatched droops of two governors enforcing a very unequal load distribution on two engines. As can be seen, identical curves are required for equal load distribution. (The last statement assumes that the engines and their fuel system characteristics are also identical.)

Low speed engines are rarely (perhaps never) found geared together in a multi-engine drive, nor are they used in the marine field as electrical generating engines, so that these applications requiring droop are not of interest here. However, use of a controllable pitch propeller requires combined control of engine speed and propeller pitch, and one way of implementing this is through intersection of engine governor droop and pitch governor droop. This scheme is discussed in Section 9.5.

Governors used by diesel engines are usually stock items* As such, they are typically proportional plus reset governors, but with a number of added features that can be used at the discretion of the engine builder. A lever added between power piston and speeder spring, this to provide the droop desired in some applications, is a common example. Fuel limiting is another feature often necessary, especially for an engine driving a controllable pitch propeller, in order to prevent the overloading that would occur if the maximum amount of fuel (that amount required for full power) were applied at low speed. The allowed travel of the power piston is therefore made a function of engine speed. Similarly, the allowed travel of the power piston is sometimes made a function of intake

---

*'Stock item' here means that the governor is not designed for the particular engine, but is one that is sold by its builder for a variety of applications.

176 CONTROL AND MONITORING

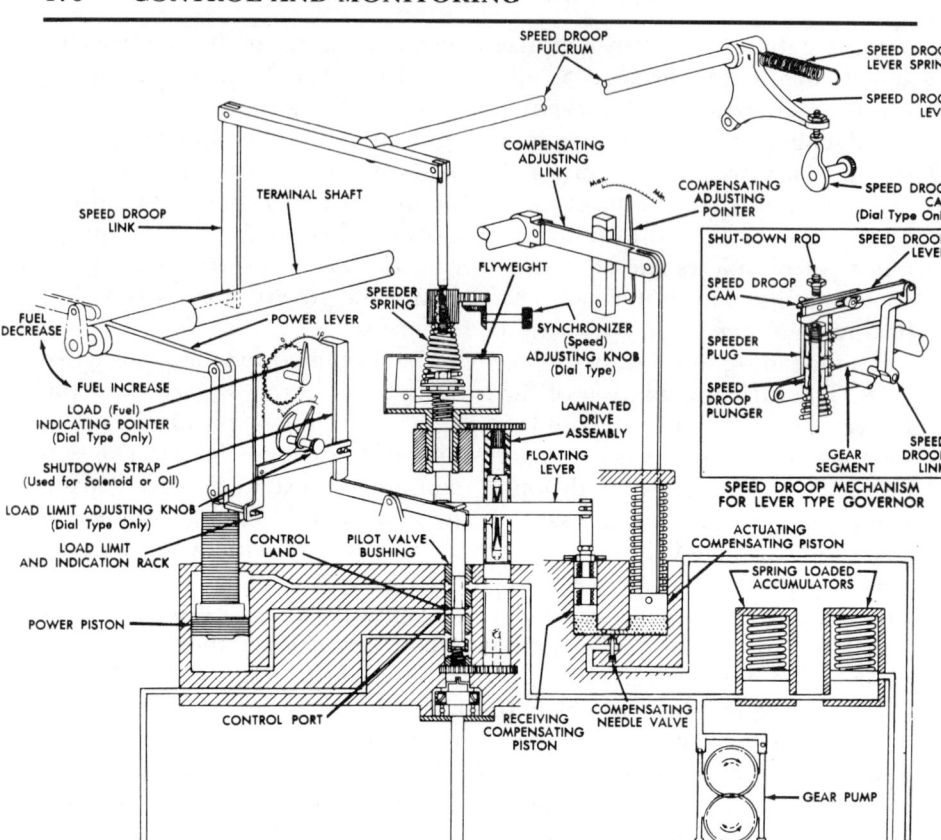

Figure 9.19 Schematic diagram to show actual governor elements [Woodward Governor Company drawing (UG40 governor)].

manifold air pressure so that the amount of fuel injected will not produce too low an air/fuel ratio. This limitation is helpful in preventing smoky combustion during engine transients.

Several other auxiliary features are possible. For example, microswitches may be actuated by the power piston tail rod so that this position can be indicated on an engine monitoring panel; a solenoid valve may impose rapid shutdown action upon receipt of an electrical signal from a remote command station; an overspeed testing device may allow the governor speed setting to be overridden briefly to test the engine's overspeed trip.

Figure 9.19 is a semi-pictorial view of a diesel governor. You will

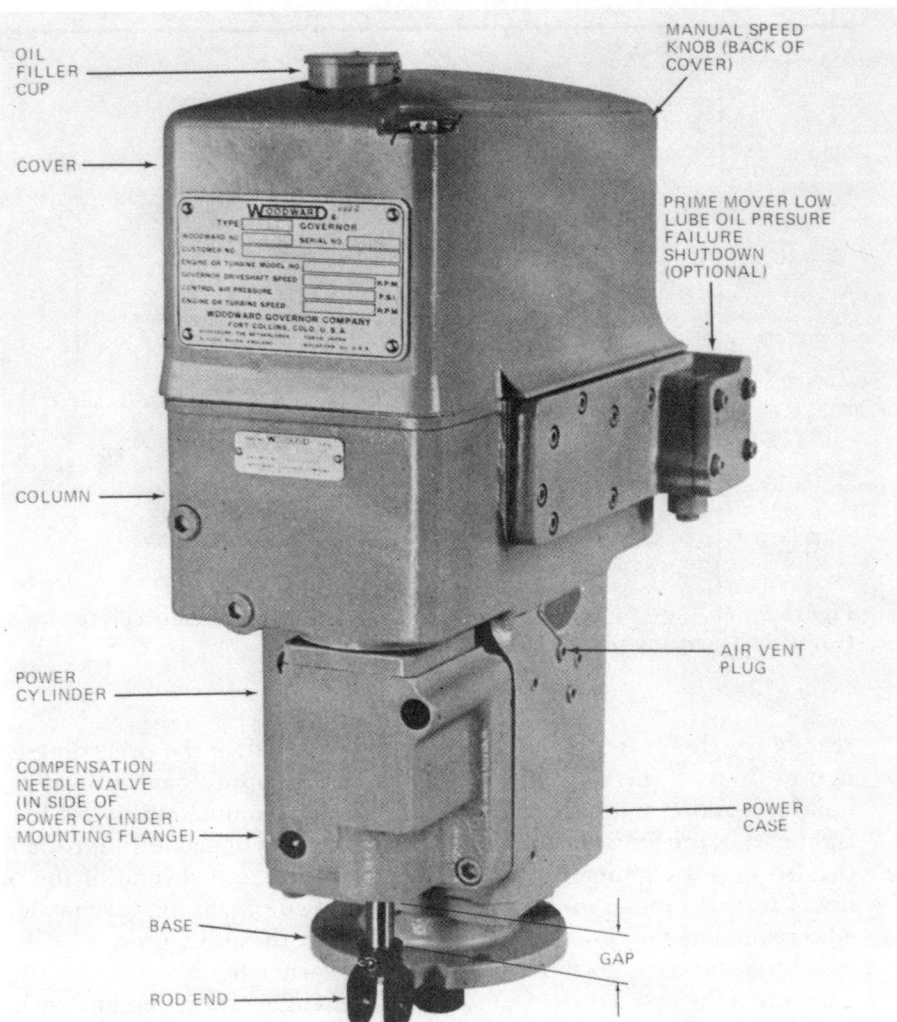

Figure 9.20 External view of an engine governor [Woodward Governor Company photograph (PGA governor)].

178 CONTROL AND MONITORING

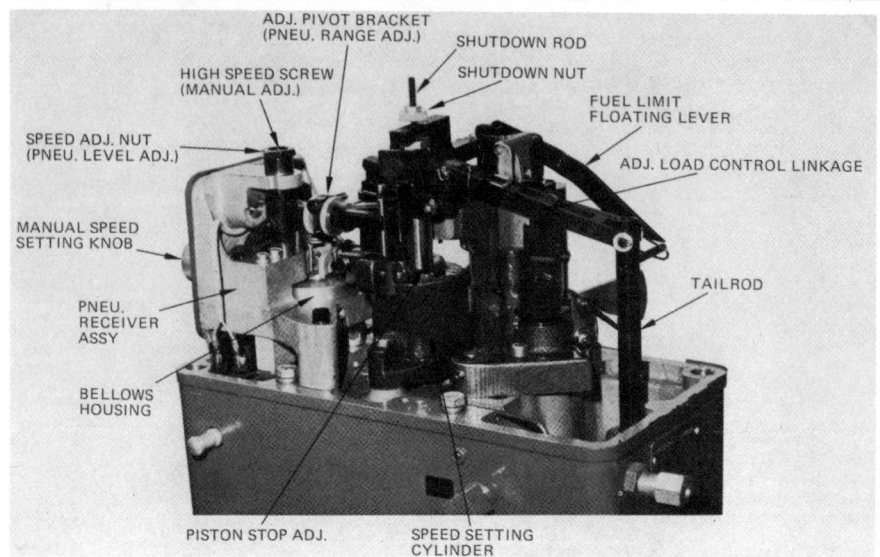

Figure 9.21 View of an engine governor with top cover removed [Woodward Governor Company photograph (PGA governor)].

readily see that it is a more complicated picture than the preceding two figures. In part, this is caused by the inevitable appurtenances needed to build a working machine; the oil pump and accumulators in the lower right part of the figure are examples of this. Means of making adjustments (to droop, for example) complicate the picture, and several of the optional features mentioned in the preceding two paragraphs are present. The reader may enjoy attempting to identify these features, as well as matching the parts to those shown in the earlier figures.

Figure 9.20 is a picture of a low speed diesel governor, and Figure 9.21 is a picture of the same unit with the cover removed.

9.9 NOTATION FOR CHAPTER 9

mm   millimeters
N rotational speed
P propeller pitch
Pa   Pascals
Q propeller torque

Chapter Ten

ENGINE AUXILIARY SYSTEMS

10.1 INTRODUCTION

Fuel and air must be supplied, and exhaust gas removed; heat must be removed from a variety of components; lube oil must be circulated externally for cooling and cleaning. And the engine must be started and controlled in its running. All of these functions demand their respective auxiliary systems. In part, the systems are integral elements of the engine. For example, the air supply compressors (turbocharger compressor, piston underside compressors, low-load auxiliary fans), air coolers, and interconnecting ducting are all components of the engine itself. But the ventilation components that supply the machinery space with air for engine consumption are not part of the engine, and not supplied by the engine builders.

The engine components are discussed in earlier chapters; the external parts to which they mate are treated here, save for engine control items, which have been covered in Chapter 9. The functional requirements (flows, temperatures, etc) of these parts must obviously be specified by the engine builders, and some of the external components (heat exchangers, for example) may be supplied by them. Nonetheless, there remains considerable work to be done by the ship designer on these supporting systems. As a minimum, the arrangement of piping, pumps, tanks, etc, is a task for this person.

180   ENGINE AUXILIARY SYSTEM

Most of the descriptions and illustrations of auxiliary systems here apply to Sulzer RND...M engines, and are borrowed from Sulzer technical literature. Where particular numerical values are helpful, they are taken from data on the 6RND90M engine (six cylinders, 2465 kW per cylinder maximum continuous output). The intention is, however, to present them as typical of all low speed engines, with the only major exception to typicality being in discussion of piston cooling: Sulzer pistons are water cooled, and while this is also true for several other engine builders, oil is used (as this is written) as piston coolant by Burmeister & Wain, and by Doxford for the lower pistons of its opposed-piston engines. Differences between the two schemes are noted at the appropriate spots.

Several important aspects of marine engineering design depend on the engine auxiliary systems. Machinery weight and electric power demand are the most prominent examples; hence some information on these points is included here.

10.2   ENGINE COOLING SYSTEMS

A large fraction of the thermal energy released by fuel combustion must be removed from the boundaries of the cylinder by liquid coolant. Additionally, lubricating oil must be cooled, and incoming combustion air is cooled to increase its density. The "heat balance" of Figure 4.3 gives a notion of energy magnitudes involved.

The heat sink for this part of the engine's unusable energy is the sea, the transfer to the sea being accomplished via a seawater circulating system in the engine room. The seawater is commonly (but not always) used directly in the combustion air cooler so that the air may be rendered as cool as possible, and is also used as coolant in the lube oil heat exchanger, but cools the engine indirectly through heat exchangers that link it to several fresh water cooling circuits.

The liquid cooling circuits typically found within the engine structure are those for the (1) jacket (cylinder liner and cylinder cover), (2) fuel injector valve, and (3) pistons. The liquid is always fresh water, with the exception of the piston coolant, which may be lubricating oil. Although water cooling of pistons is used by several engine builders because of its excellent heat transfer and heat transport properties, oil is used by others because its possible leakage does not entail a threat of contamination to the engine lubricant. (Recall from Chapter 3 the means by which coolant must be carried to and from the moving pistons.) Seawater is not used directly in any circuit within the engine because its incoming temperature is awkward to control, and because of the dangers of corrosion and

ENGINE COOLING SYSTEMS 181

deposition of sea salts within the engine. The fresh water coolant temperature is regulated by controlling the flow of the seawater, and its corrosion and deposition behavior is inhibited by additives, and readily so because the coolant circuits are closed loops.

The individual cooling circuits just listed are used in lieu of a single circuit because the several services have differing temperature and pressure specifications, and in one instance—the piston cooling circuit—there is the possibility of oil in-leakage that must be isolated from as much of the other parts of the cooling system as possible.

Diagrams and descriptions follow for the cooling systems used by one engine builder; recall that Sulzer data are being used for illustration, and that numerical values apply to Sulzer's 6RND90M engine, unless stated otherwise.

A. Fuel Injector Cooling

Water is circulated in parallel through passages surrounding the injector, and through an expansion tank of approximately 600 liters volume. No exchange with seawater is required here because dissipation to the air from pipe and tank surfaces is sufficient cooling. On the other hand, a steam heating coil is fitted into the expansion tank so that the injectors can be preheated before engine start-up. Required pump capacity is 4.8 m^3/h. Two such pumps are installed; one is standby.

B. Jacket Water Cooling

This system cools the cylinder liners, cylinder covers, and turbocharger compressors. It is shown schematically by Figure 10.1 (Sulzer dwg 2-107.095.135-1). Water is circulated from the seawater heat exchanger (item 04) by the pumps to the engine jacket inlet pipe, thence via an air separator back to the heat exchanger. Upon leaving the engine, part of the water is diverted through the turbocharger. Two pumps (one standby) of 270 m^3/h are installed. The heat exchanger design conditions require cooling from 68 to 54.2 C, an exchange of 4330 kW, which is approximately 11 percent of the fuel energy input. Each jacket water pump requires an energy input of about 30 kW. Most of this energy appears as heat in the water, and so contributes to this extent to the 4330 kW heat exchange.

The system also features an expansion tank and a steam heater for preheating water before engine start-up.

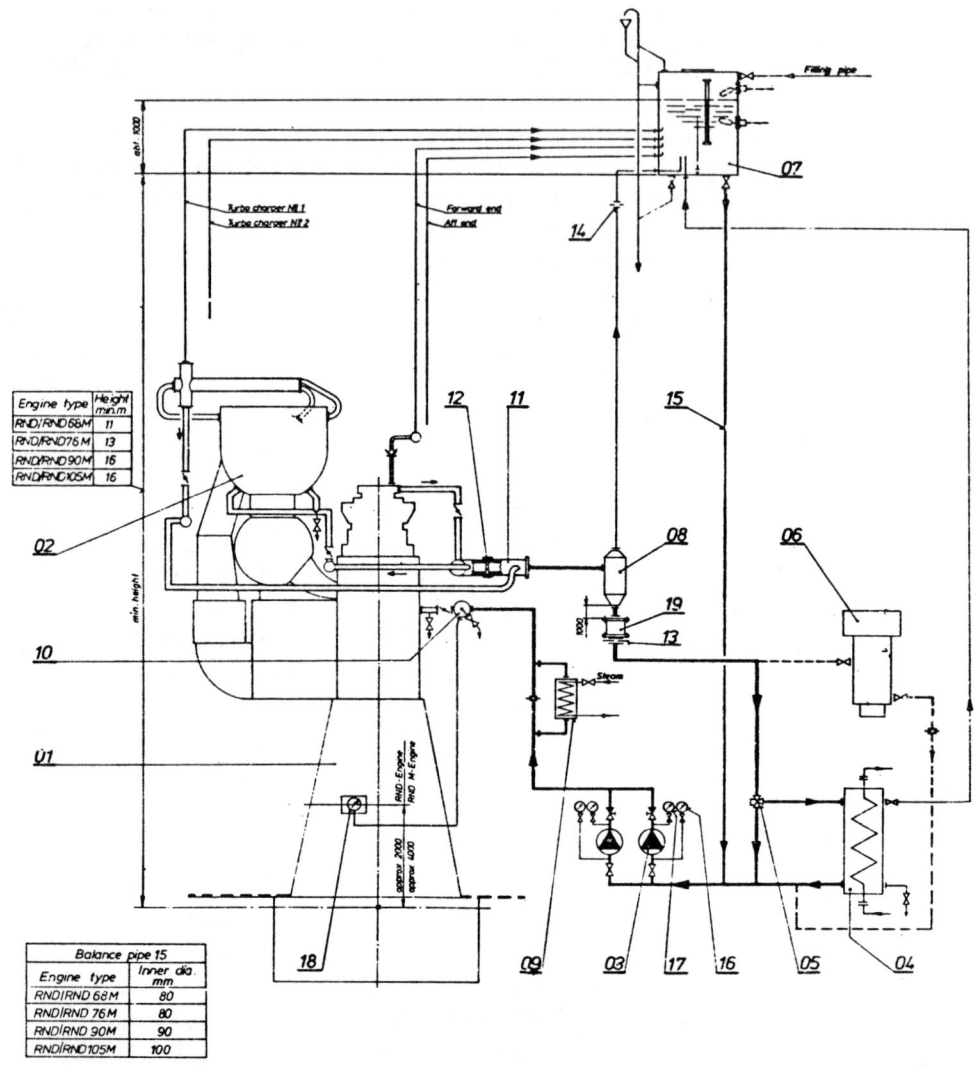

Figure 10.1 Jacket water cooling system for Sulzer RND..M engines (Sulzer drawing).

01 Main engine
02 Turbocharger
03 Jacket F.W. pump
04 Jacket F.W. cooler
05 Automatic temperature control valve
06 Fresh water generator
07 Expansion tank—with low level alarm
08 Air separator (centrifugal type)
09 Heater
10 Jacket F.W. inlet pipe

The figure shows a vacuum distilling plant ("fresh water generator," item 06), this being an optional addition to the system that uses the engine's rejected thermal energy to produce fresh water. The figure shows the unit piped in parallel with the heat exchanger, with the note that not more than 30 percent of the heat to be transferred shall be extracted by it. If more than 30 percent can be absorbed by the water generator, a series arrangement is used, and the necessary piping and control features are described in the Sulzer literature. The details are not to be given here, but we note that the vital principle in any installation of a heat-using device in the cooling system is to avoid interference with the cooling function—neither retard cooling of the water nor overcool it. As an example of possible overcooling, picture that a high demand for fresh water distillation might occur while the engine is lightly loaded. In such a situation the engine water could be cooled below the 54.2 C outlet temperature specified by Sulzer. Control valves must therefore be arranged so that engine demand overrides that of the distiller.

Three possible distiller arrangements are shown in Figure 10.2 (Sulzer dwg 2-107.095.935). The first (leftmost) is the same as that indicated in Figure 10.1.

C. Piston Cooling System

This system cools the pistons via telescoping pipes (recall Figure 3.4). It is shown schematically by Figure 10.3 (Sulzer dwg 2-107.095.945), and features pumps, heat exchanger, steam preheating coil, and expansion tank, all similar to the jacket water system. Pump capacity is 66 m^3/h per pump, with one pump being a standby. The heat exchanger cools the water from 61.8 to 45 C, for a heat exchange of 1285 kW, which is

11 Jacket F.W. outlet pipe
12 Throttling piece for adjustment of turbocharger cooling water quantity
13 Throttling disk or adjustable throttling piece for adjustment of system cooling water pressure
14 Throttling piece orifice in air vent pipe
15 Balance pipe
16 Pressure gauge on F.W. pump suction side
17 Pressure gauge on F.W. pump pressure side
18 Pressure gauge for jacket fresh water
19 Fluid flow stabilizer

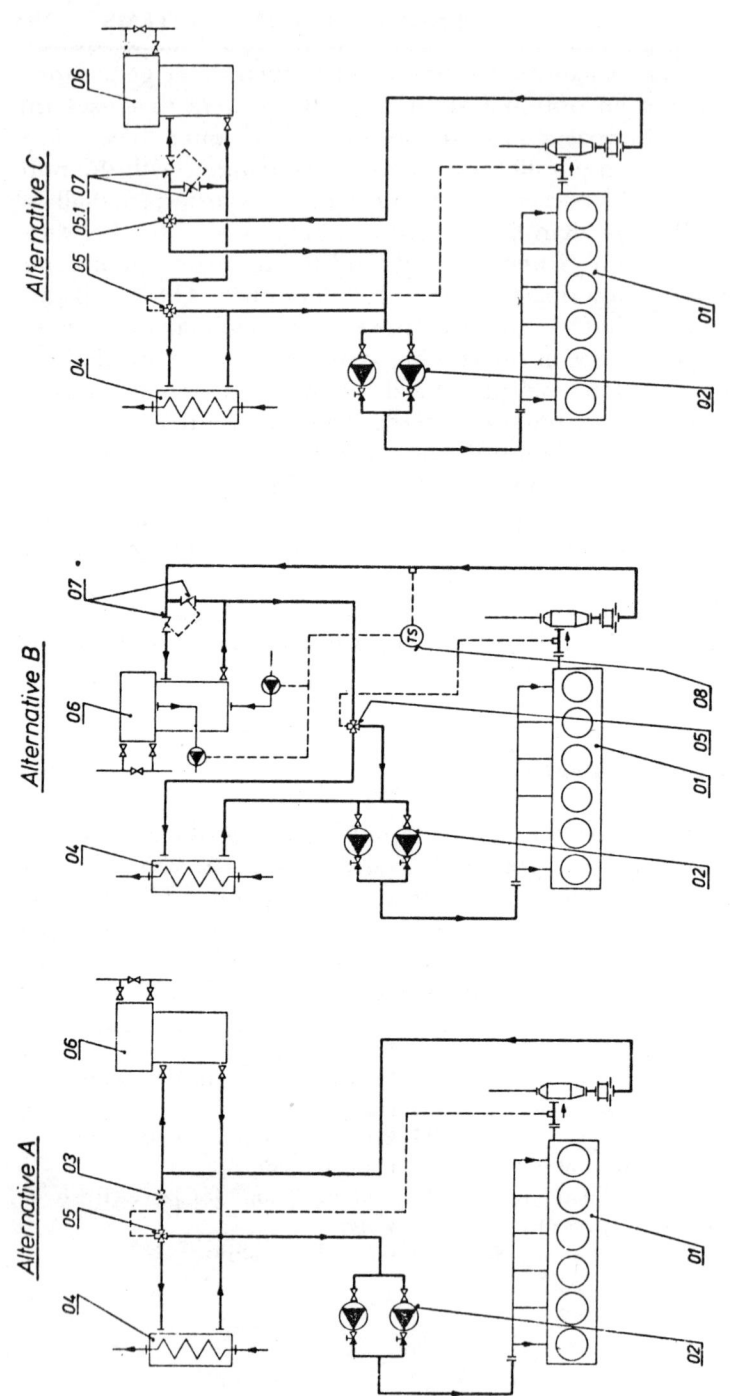

ENGINE COOLING SYSTEMS   185

approximately 3 percent of the fuel energy input. Each piston cooling pump requires an energy input of about 15 kW.

Some of the piston cooling water inevitably leaks from the sliding joints of the piston telescopic pipes, especially after their glands become worn. The leakage water is likely to be contaminated with oil, and hence must not be re-used for cooling unless the oil is removed. One alternative is to discard the contaminated water, but the resulting makeup rate may be a nuisance (Sulzer mentions a rate of at least 20 liters/hr per cylinder for worn glands), and re-use after purification is the recommended (according to Sulzer) alternative. Several purification schemes are possible (for example, gravity separation, centrifugation). None are shown in Figure 10.3, nor are the piping and pumps necessary for collection of the leakage shown. The sump tank shown in the figure is the point of return for the purified water, and can serve as an inspection point (via a sight glass) for condition of the water.

The coolers mentioned above, as well as the combustion air cooler and the lube oil cooler, transfer their heat to seawater, usually via a once-through seawater system. Figure 10.4 (Sulzer dwg 2-107.095.090) shows a typical system diagram, with rated temperatures in and out of all heat excangers noted. Pump capacity is 792 m³/h per pump, with one of the two pumps being a standby. The sum of the rated heat transfer capabilities of the four heat exchangers served is 11,011 kW, or approximately 28 percent of the fuel energy input. Note that a distilling plant (fresh water generator) is also shown, since such a unit requires cooling water for condensing the vapor it produces.

An alternative to the more traditional cooling system is the "central cooling system" that uses an intermediate fresh water loop to cool the lube oil, piston water, and combustion air. This loop in turn transfers its heat to seawater via a "central cooler." The principal advantage is easier control of water temperature, especially that of the water used to cool the combustion air. Excessive cooling here, as from over-cold seawater, may cause condensation of atmospheric moisture within the cylinder, thus exacerbating the problems of cylinder corrosion. The obvious disadvantage is additional components, particularly pumps for circulating the intermediate water, and the additional heat exchanger.

Figure 10.2 Methods of connecting fresh water generators to the cooling systems of Sulzer RND..M engines (Sulzer drawing).

01 Main engine
02 Jacket cooling water pump
03 Valve for flow regulation
    (normally not necessary)
04 Jacket cooling water cooler

05 + 05.1 Automatic temperature
          control valve
06 Fresh water generator
07 Butterfly valve
08 Low-temperature cutout switch

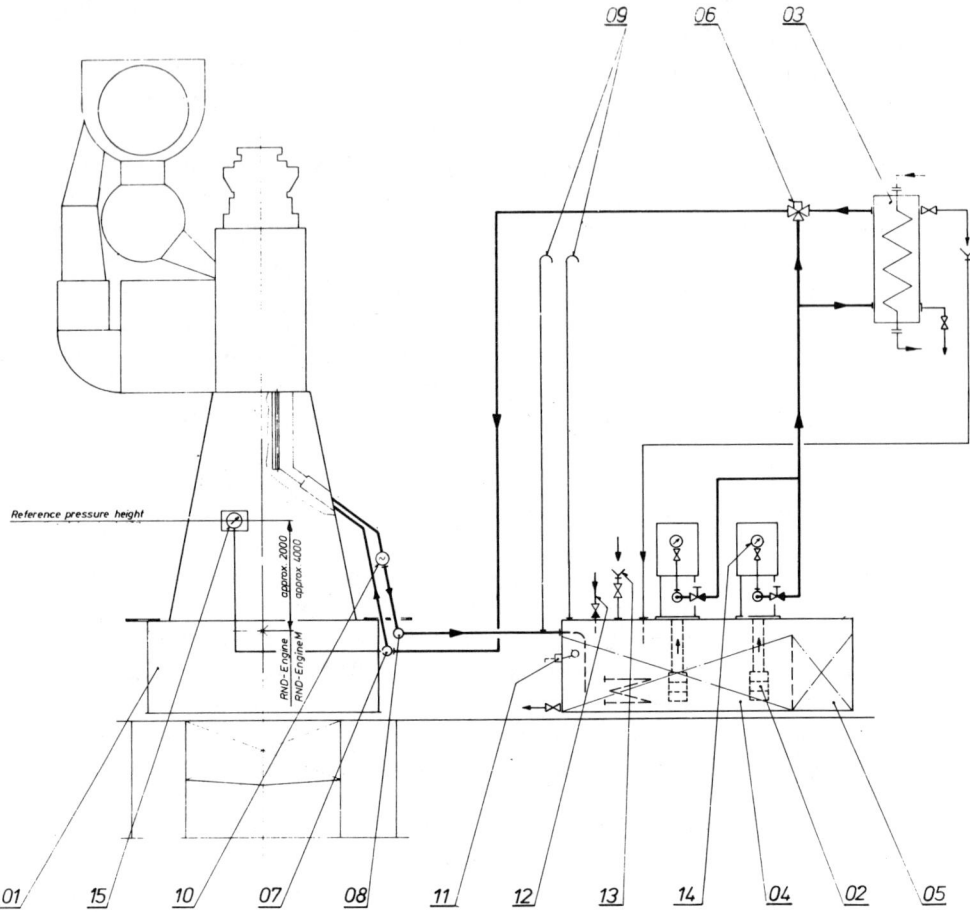

Figure 10.3 Piston cooling for Sulzer RND..M engines (Sulzer drawing).
01 Main engine
02 Piston cooling water pump
03 Piston cooling water cooler
04 Piston cooling water drain tank
05 Piston cooling leakage water tank
06 Automatic temperature control valve
07 Piston cooling water inlet pipe
08 Piston cooling water outlet pipe
09 Air vent pipe
10 Sight glass
11 Low level alarm
12 Filling pipe
13 Chemical treatment inlet
14 Pressure gauge on C.W. pump (press. side)
15 Pressure gauge for piston cooling water

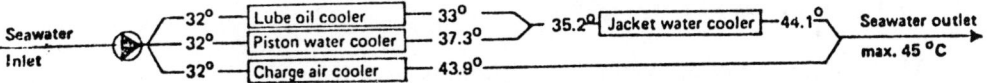

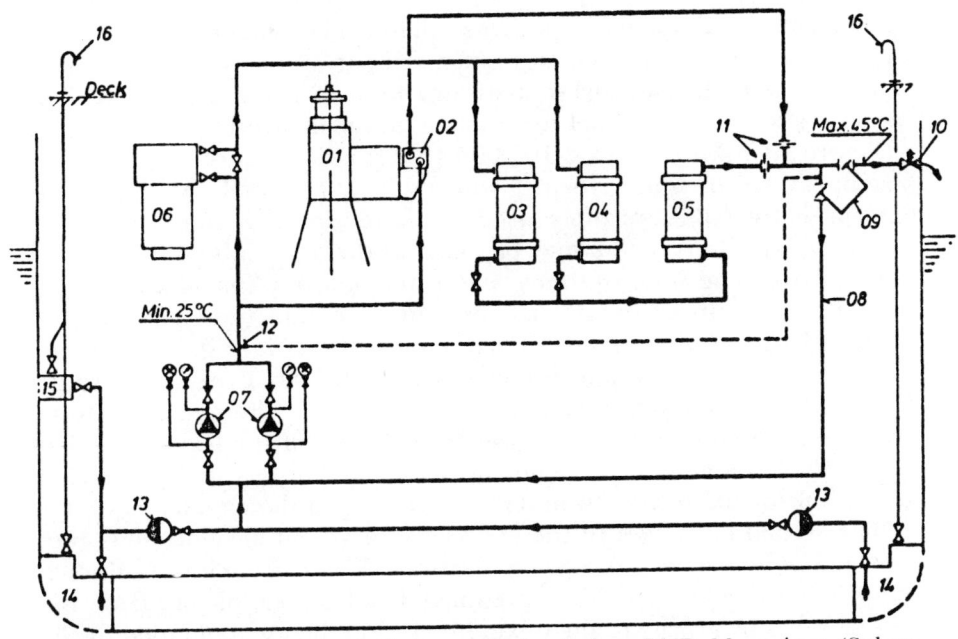

Figure 10.4 Seawater circulating system for Sulzer RND..M engines (Sulzer drawing).

01 Main engine
02 Air cooler
03 Lubricating oil cooler
04 Piston F.W. cooler
05 Jacket F.W. cooler
06 Fresh water generator
07 Seawater pump
08 Warm seawater return line
09 Automatic temperature control valve (dual butterfly type)
10 Overboard n.r. valve (spring loaded)
11 Throttling disk
12 Temperature feeler
13 Seawater filter
14 Lower sea chest
15 Upper sea chest
16 Air vent

188 ENGINE AUXILIARY SYSTEM

10.3 FUEL SYSTEM

The general function of the fuel system is to deliver fuel from the ship's tanks to the injection pump on the engine. Nominally, then, only a pipe and a pump are required, but the additional function of conditioning the fuel demands additional components. All low speed engines burn "heavy oil," which must be heated to reduce viscosity, and treated to reduce impurity content. The fuel system also usually includes provision for switching to a distillate fuel for starting and maneuvering.

Figure 10.5 (Sulzer dwg 2-107.095.150) is a diagram of a fuel system recommended by Sulzer for use with its RND and RND..M engines, a system that is typical of those used by all low speed engines.

The heart of the system may be taken as the array of tanks (items 07 through 10). The *heavy oil settling tank* is the first stop for oil transferred from the ship's fuel storage tanks by transfer pumps (not shown). This tank is sized to hold at least a 1 day supply of oil; during the resulting residence time gross impurities, especially water, are expected to settle toward the bottom of the tank, whence they may be drained away. Note that the diagram indicates a suction level for the oil at a level somewhat above the bottom, this to avoid the settled impurities when the oil is drawn off for the next stop in its passage toward the engine.

The settled oil passes to the *day tank* (09) via purification units (not shown, but note lines "to heavy oil separator" and "from heavy oil separator"). This tank also holds approximately a 1 day supply of fuel. From here it flows by gravity to the mixing tank, and thence is pumped by the fuel booster pumps to the engine.

The *mixing tank* is comparatively small, perhaps 1 or 2 m^3 in volume (the figure is obviously not to scale), and serves as the entry point for distillate fuel when it is used. The distillate ("diesel oil") is used in lieu of the normal heavy oil at times, such as low power operation, when sufficient steam* for heavy oil heating may not be available, and in maneuvering before shutdown in port, this so that the engine internals will be suitably clean for any maintenance work. Use of the mixing tank lessens the abruptness of the switchover from one fuel to another, particularly with respect to temperature, which can be quite different for the two different oils. Except during a changeover interval, the two fuels are not used simultaneously.

The *diesel oil day tank* serves the same function as the heavy oil day tank,

---

*Such steam is usually produced by the engine exhaust in a 'waste heat boiler." At low engine loads, both quantity and temperature of exhaust gas are reduced, thus sharply limiting steam production.

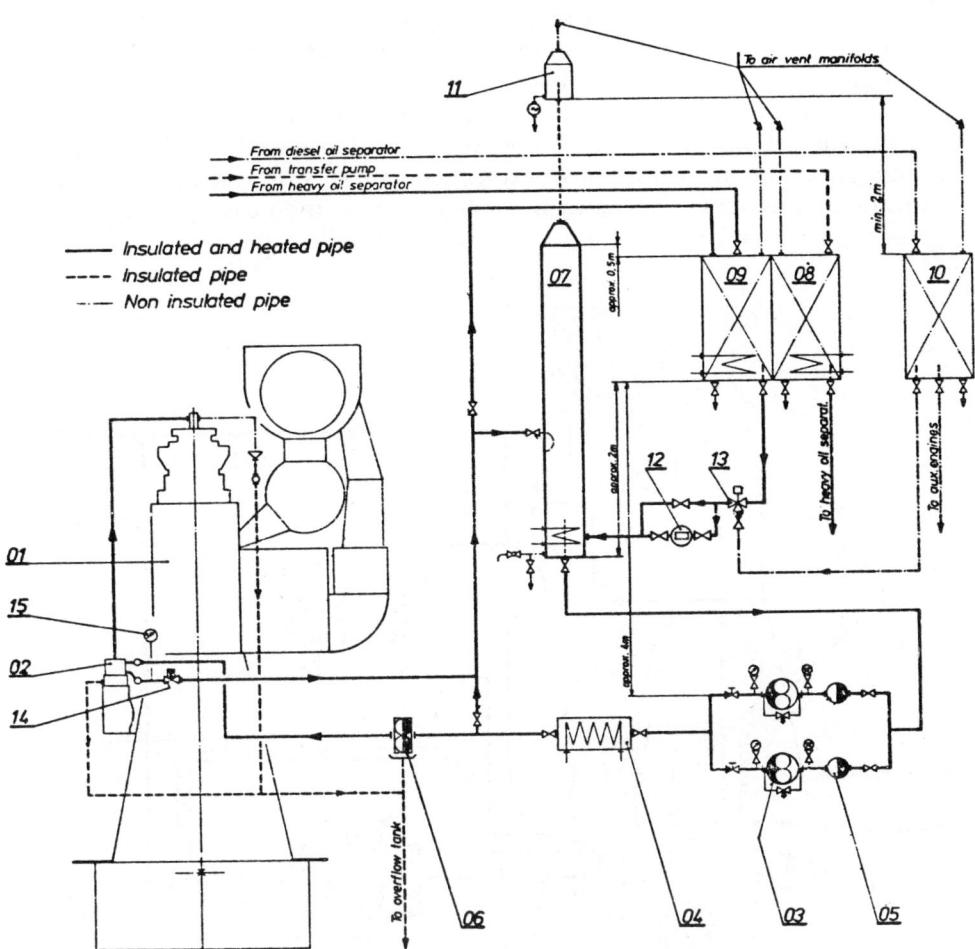

Figure 10.5 Fuel system for Sulzer RND..M engines (Sulzer drawing).
01 Main engine
02 Fuel injection pump on main engine
03 Fuel oil booster pump
04 Steam fuel end-heater
05 Suction filter, heatable
06 Duplex filter, heatable
07 Mixing tank, heatable and insulated
08 Heavy oil settling tank, heatable
09 Heavy oil daily tank, heatable
10 Diesel oil daily tank
11 Condensation water trap
12 Fuel oil flow meter
13 Three-way valve, pneumatic or electric operated
14 Spring-loaded adjustable relief valve on main engine
15 Pressure gauge

and is approximately the same size as that tank. However, the diesel day tank also serves all auxiliary (generator drive, usually) diesel engines, so that the designer in sizing the tank must take into account the fuel demands of these engines as well.

Each of the two fuel booster pumps (one running, one standby) has a capacity of approximately 2.5 times the fuel consumption of the engine at maximum continuous output. The surplus oil normally returns to the mixing tank; alternatively it can return to the heavy oil day tank.

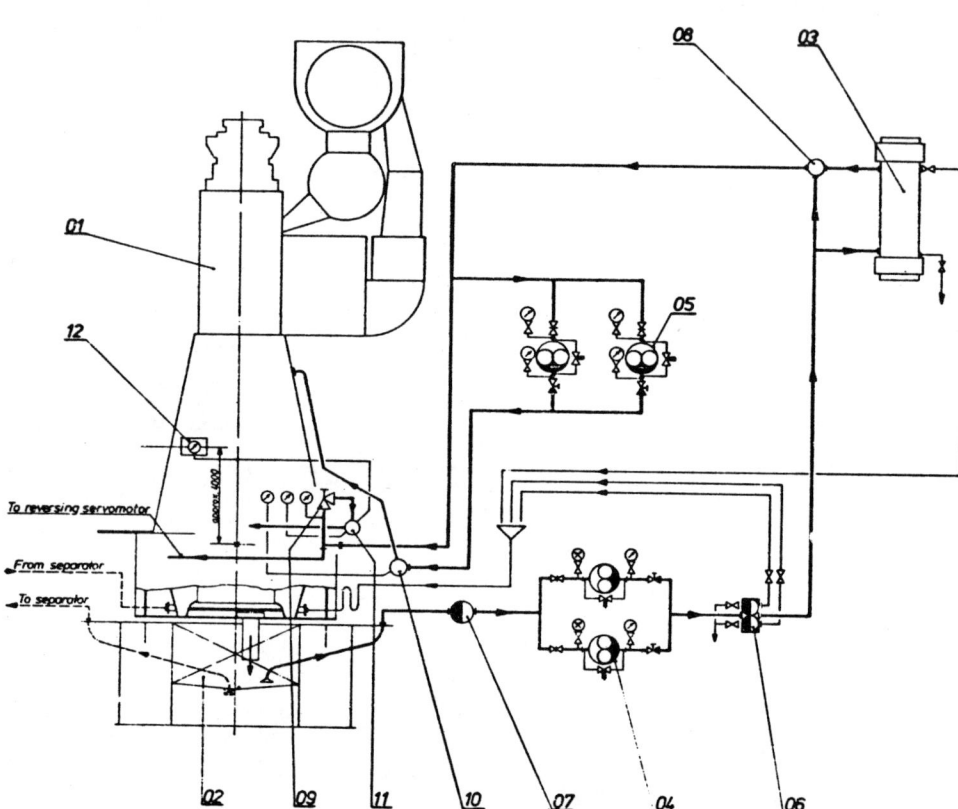

Figure 10.6 Lubricating system for Sulzer RND..M engines (Sulzer drawing).
01 Main engine
02 Oil drain tank
03 Lubricating oil cooler
04 Lubricating oil pump
05 Crosshead lubricating oil pump
06 Lubricating oil filter
07 Suction oil filter
08 Automatic temperature control valve
09 Pressure reduction valve
10 Crosshead lubricating oil inlet
11 Bearing lubricating oil inlet pipe
12 Pressure gauge for bearing lubricating oil

Observe that all tanks serving heavy oil have steam heating coils to reduce viscosity for good pumpability. A heater (04) downstream of the booster pumps raises the temperature to produce the still-lower viscosity needed by the engine for atomization. The temperature reached depends on the "heaviness" (that is, the viscosity at a standard temperature) of the oil being used, but typically is in the range 100 to 120 C.

The purification for the heavy oil, taking place between settling tank and day tank, can be accomplished in several ways. The ones commonly available for use are discussed in Chapter 8.

Gas fuel—available from the "boil off" of LNG ships—has been discussed in Chapters 8 and 9. A gas fuel system and its control interactions with the normal oil system are pictured by Figure 9.15. Recall that the gas enters the engine cylinder via a gas injection valve, and does not mix with the oil, nor share common components outside the engine cylinder. Figure 10.5 would therefore not be different if the engine were also equipped for gas burning.

10.4 LUBRICATING OIL SYSTEMS

We speak of "systems"—plural—because there are two. Recall from Chapter 8 that cylinder walls require a lubricant of properties different from those of the bearing lubricant. Separate systems must therefore be installed. Bearing lubrication is supplied by a closed system, with cooling and purification for continuous reuse. Cylinder lubricant is consumed within the cylinder, and hence is supplied by a once-through system. Figures 10.6 and 10.7 (Sulzer dwgs 2-107.095.287 and 4-107.095.065, respectively) show typical versions of these systems.

The Sulzer engines require bearing oil supply at three distinctly different pressures: (1) 8 to 16×10^5 Pa for the crosshead bearings, (2) 1.5 to 2.5×10^5 Pa for all other bearings, and (3) 3 to 4×10^5 Pa for the hydraulic components of the engine control system. The provision of these pressures can be observed in Figure 10.6: the main lube oil supply pumps furnish the medium pressure and the low pressure via a reducing valve; the high pressure is furnished by crosshead lube oil pumps that are in series with the main pumps.

The source of the recirculated lube oil is a sump tank below the engine room tank top, into which oil from the engine drains continuously. The tank is sized to hold all of the oil filling the remainder of the system, plus enough for a safe coverage of the suction pipe when the system is in operation. The function of the pumps is to circulate the oil through the cooler, and to repressurize it to the levels just mentioned. Capacity of each pump (for the 6RND90M engine) is 180 m³/h; capacity of each of

192 ENGINE AUXILIARY SYSTEM

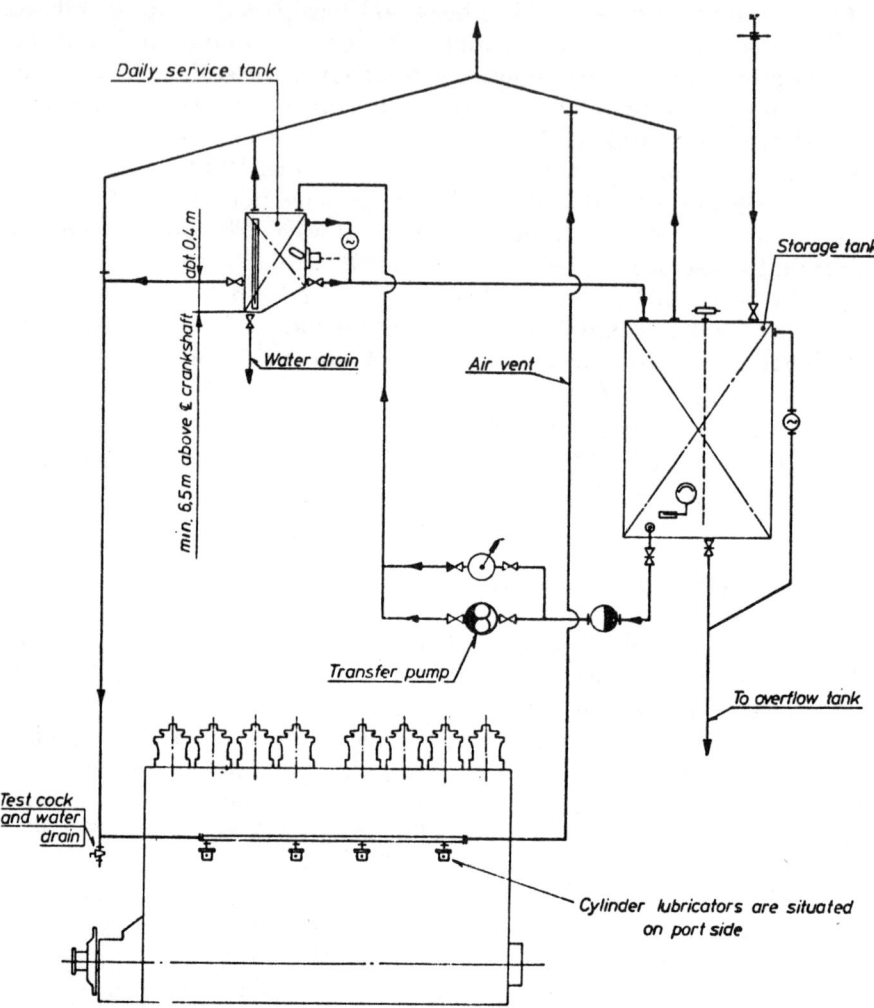

Figure 10.7 Lubricating system for Sulzer RND..M engines—cylinder lubrication (Sulzer drawing).

the crosshead pumps is 45 m³/h. In both cases, one pump in each pair is normally in use, the other is standby.

The 6RND90M oil cooler dissipates 246 kW, or approximately 0.6 percent of the fuel energy input, cooling the oil from 45.7 to 43 C. Each main pump requires an energy input of about 35 kW, and each crosshead pump about 30 kW. Roughly one-quarter of the heat dissipated therefore originates as pump input.

Engines that cool their pistons with oil do so with the bearing oil

LUBRICATING OIL SYSTEMS 193

system, and that system is essentially unchanged from that described here, except for the much greater heat dissipation. The heat described in Section 10.3 as being transferred by the piston cooling water must instead be transferred by the lube oil cooler.

A purification system removes oil from the sump tank and returns it after purification, usually by a centrifuge. Figure 10.8 is a diagram of a typical system, this one fashioned around an Alfa-Laval centrifugal purifier. (The lube oil circulating system is also included in this figure; note the right-hand part.) Water, sludge, and solid particles are removed centrifugally. The flow of oil to the purifier is smal compared to that of the main system, and energy exchanges are likewise comparatively small. Flow rate (as recommended by Alfa-Laval) is sufficient to treat the entire system volume two or three times during a day if the oil is straight

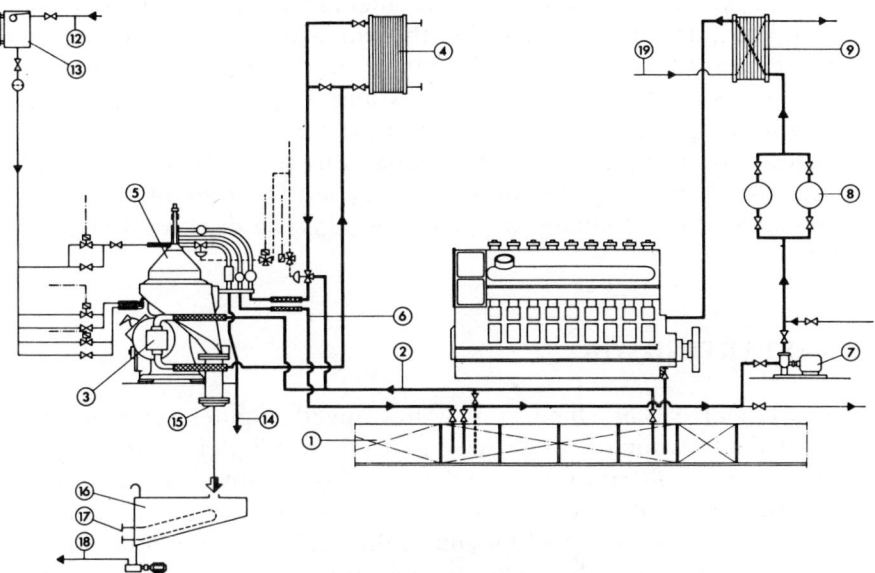

Figure 10.8 Lubricating oil purification system (Alfa-Laval AB).
1 Sump tank for dirty oil from engine
2 Dirty oil to purifier
3 Pump
4 Heater
5 Alfa-Laval purifier
6 Purified oil
7 Pump for lubricating oil
8 Filter for lubricating oil
9 Cooler for lubricating oil
12 Cold fresh water
13 Operating water tank
14 Water outlet
15 Sludge outlet
16 Sludge tank
17 Heating coils
18 From sludge pump
19 Cooling water inlet

mineral, or up to five times if it is a detergent oil. The heater raises oil temperature to promote impurity separation through lower oil viscosity. The water tank supplies sealing water to the purifier.

Figure 10.7 shows the Sulzer cylinder oil system, and rather simple it is since the only function is to move oil from a supply tank to the engine cylinder lubricators. The storage tank is filled from a deck connection. Oil is transferred to the day tank by a small pump, and thence flows by gravity to the engine. Day tank capacity is about a two-day supply, or about 0.6 m^3 for the 6RND90M engine. The storage tank size depends, of course, on the time between fillings expected by the operators; about 100 days might be typical.

Sulzer's specified cylinder oil consumption rate is 0.7 to 1.0 g/kWh. A rate for this consumption is always cited in literature relating to diesel propulsion economics, since the cost of oil consumed is a significant operating expense (for example, see [Femenia (1970)]. The figures found in such literature are usually within the range stated by Sulzer, or even less; the diesel bulletin of the Society of Naval Architects and Marine Engineers [Society of Naval Architects and Marine Engineers (1975)] advises a design estimating rate of about 0.6 g/kWh.

Sulzer states its bearing oil consumption rate to be about 0.07 g/kWh. Higher rates than this are sometimes advised in other literature. For instance, the diesel bulletin just mentioned gives a design estimate rate of 0.3 g/kWh.

10.5 · STARTING AIR

It is noted in earlier chapters that low speed engines are started by controlled admission of compressed air to their cylinders, and that starting air valves are fitted to the cylinders for this purpose. Timing of these valves and distribution of air to them are two of the functions of the engine control system. The function of the auxiliary air system is simply to compress and store the necessary air, and to convey it to the engine.

Figure 10.9 (Sulzer dwg 2-107.095.101) diagrams a typical starting air system.

The principal parameters of interest are the capacity of the air storage tanks and the pressure that must be maintained. (Storage of air is necessary because of the impracticality of providing an air compressor big enough to do the job on demand.) These depend on the quantity and pressure requirements of the engine for one start, and upon the requirements of regulatory bodies for a minimum number of starts to be accomplished without renewing the air supply. The American Bureau of Shipping [American Bureau of Shipping (1980)], for example, requires 2 air

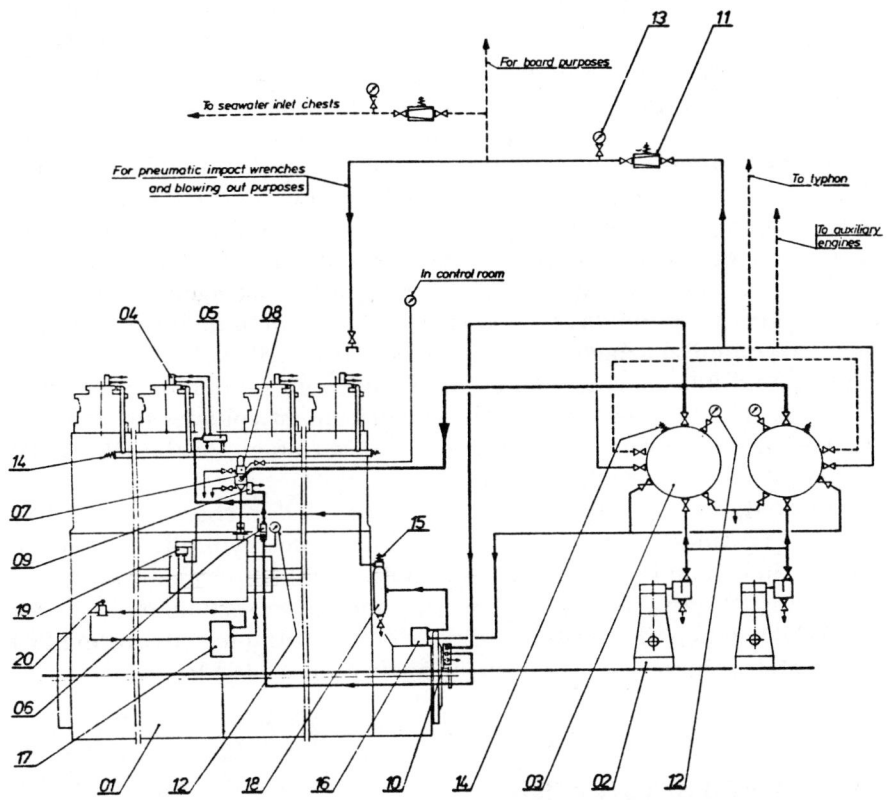

Figure 10.9 Starting air system for Sulzer RND..M engines (Sulzer drawing).
01 Main engine
02 Air compressor
03 Starting air receiver
04 Starting valve on cylinder cover
05 Starting air distributor
06 Starting air valve (pneumatic and manual)
07 Automatic starting air stop valve
08 Nonreturn valve
09 Control valve
10 Interlock valve on turning gear
11 Automatic reducing valve 30–7 bar
12 Pressure gauge 30 bar
13 Pressure gauge 7 bar
14 Safety valve 30 bar
15 Safety valve 7 bar
16 Pressure reducing and filtering unit
17 Pneumatic logic box
18 Control air bottle
19 Main blocking valve for control air
20 Starting button

tanks with a total capacity sufficient for at least 12 starts. The air compressor must be capable of charging these tanks to the required pressure within one hour. In addition, an emergency compressor whose driver does not itself depend on the air supply for starting must be provided. Other regulatory bodies differ in the number of starts required, and other bodies may allow a reduction if a controllable pitch propeller is used.*

The minimum pressure required varies among different engines. For Sulzer engines the minimum pressure is generally 7×10^5 Pa, but 10×10^5 is specified for the four-cylinder RND68 and RND76 engines.

The air quantity per start also depends on the particular engine, and on the pressure in the tank when the start is attempted. Sulzer's recommended volume at a maximum storage pressure of 30×10^5 Pa is roughly 2.3 engine displacement volumes for engines with a minimum number of cylinders (4 to 6), down to about 1.5 displacement volumes for 12-cylinder engines. For example, the recommended volume for the 6RND90M engine is 6.8 m³ in each of the two tanks.

As noted, the capacity of compressors installed to charge the starting air tanks is also governed by regulatory body rules, typically rules based on the time required to charge the tanks for the specified number of starts. For 12 starts charged in one hour, the Sulzer recommendation for the 6RND90M engine is two compressors (both running) delivering 204 m³/h (atmospheric) per compressor against 30×10^5 Pa.

10.6 VENTILATION AIR

Outside air must be supplied to machinery spaces in order to meet the demands of all engines and auxiliary boilers for combustion air and the demand of air compressors, and to maintain a reasonable ambient temperature within the spaces.

Air consumptions of the engines are readily estimated from the known air/fuel ratios, or they may be stated directly by the manufacturers. For instance, the Sulzer 6RND90M engine at 100 percent load requires 140,700 kg/h at 45 C, 60 percent relative humidity, which is approximately equivalent to an air/fuel ratio of 42/1. Auxiliary engines may require a lesser ratio, especially if they are four-stroke engines. For example, a Caterpillar D399TA engine (875 kW) has an air/fuel ratio of about 24/1.

Auxiliary oil-fired boilers require air roughly equal in mass to the steam

---

*Recall that a controllable pitch propeller obviates the need to stop and start the engine during maneuvering.

VENTILATION AIR

produced, that is, production of 1.0 kg/h of steam requires about 1 kg/h of combustion air.

Air compressor consumptions are the same as their capacity ratings.

Of course, all air consumers are not likely to be running simultaneously; if the propulsion engine is operating at or near its rated power, it may be providing all auxiliary energy via its waste heat boiler. If so, then neither oil-fired boilers nor auxiliary diesel generators will be operating.

Heat dissipated from the surfaces of the engine and other components must be removed by air circulated through the machinery spaces. Some heat is picked up by the combustion air before it is consumed, but the balance has to be carried away by additional air that is supplied, then discharged to atmosphere. A detailed calculation of the air required is certainly possible, but cumbersome and not likely to have any accuracy because of the many uncertainties handicapping calculations of heat flows from the multitude of hot surfaces in the machinery spaces. Estimating rules therefore must suffice in the usual design circumstances. Several are noted following; all give the total requirement (combustion air plus ventilation air).

1. *Change of air per unit time* The gross machinery space volume (gross = no deduction for volume occupied by machinery) multiplied by a constant gives volume supplied per hour. Sulzer recommends 30 as an appropriate value for the constant.

2. *Ratio to combustion air supply rate* The combustion air supply rate multiplied by a constant gives the total supply rate. Sulzer states that the constant should lie between 1.5 and 2.0.

3. *Temperature rise calculation based on engine heat loss* The energy input to the main engine multiplied by a constant gives an estimate of the heat dissipation that must be removed. Sulzer gives this constant as approximately 2.3 percent of the energy input, or about 220 kJ/kWh (where the kW is the engine rated power). The product of air mass flow rate, air specific heat, and air temperature rise between supply and exhaust then equals 220 multiplied by engine power. Since two air masses (combustion and ventilation) are actually involved, the sum should be split into two parts, as expressed by the following formula:

$$(\dot{m}_c \Delta T_c + \dot{m}_v \Delta T_v) C_p = 220P \qquad (10.1)$$

P = engine power in kW
C_p = specific heat of air, kJ/kgC

198 ENGINE AUXILIARY SYSTEM

$\dot{m}_c$ = mass flow rate of combustion air, kg/h
$\dot{m}_v$ = mass flow rate of ventilation air
ΔT = combustion air temperature rise before consumption, C
ΔT = ventilation air temperature rise

Reasonable values for ΔTc and ΔTv can be chosen (Sulzer lists 8 and 15 C, respectively), $\dot{m}_c$ is calculated previously, and C_p is known from handbook data; $\dot{m}_v$ can thus be calculated from this formula and added to mc to give the total air requirement.

Figure 10.10 (Sulzer dwg 1-107.095.254) illustrates a possible arrangement of air supply to an engine room. The general principle followed is delivery of air close to the points of consumption, and to points near major heat sources. Delivery close to points of consumption reduces the possibility of the air picking up oily vapor, and so reduces oily fouling of intake filters and compressor surfaces.

Combustion air to the propulsion engine can also be supplied via ducts that connect to the turbocharger compressor inlets. In such an alternative the designer must take care that neither thermal expansion of this ducting nor its weight place large forces and moments on the turbocharger, that rain or spray cannot enter the turbocharger via this path, that pressure drop between atmosphere and turbocharger is not excessive (the engine builder will specify a maximum allowed), and that the duct does not serve as a pathway for engine noise to the outside world. Expansion joints, baffles, drains, and sound absorbing linings or baffles are thus likely to be required.

10.7 EXHAUST

Exhaust ducting is apparent in several figures, most recently Figure 10.-10. It should be evident that in terms of bulk, the exhaust system is the largest of the auxiliary systems. In addition to the ducting, it usually includes a waste heat boiler, followed by a spark arrester. If the waste heat boiler is not fitted, a silencer must be provided to effect the sound dampening that otherwise would occur in the boiler.

The function of the exhaust duct is obviously to provide a path to atmosphere; an essential element of this provision is that it be done without excessive pressure drop from the turbocharger outlet to atmosphere. A minor consideration here is that the engine is furnishing the energy to overcome the resistance fo the duct, hence is burning fuel for that purpose that otherwise would contribute to propulsion. The major consideration is avoidance of high exhaust temperature that accompa-

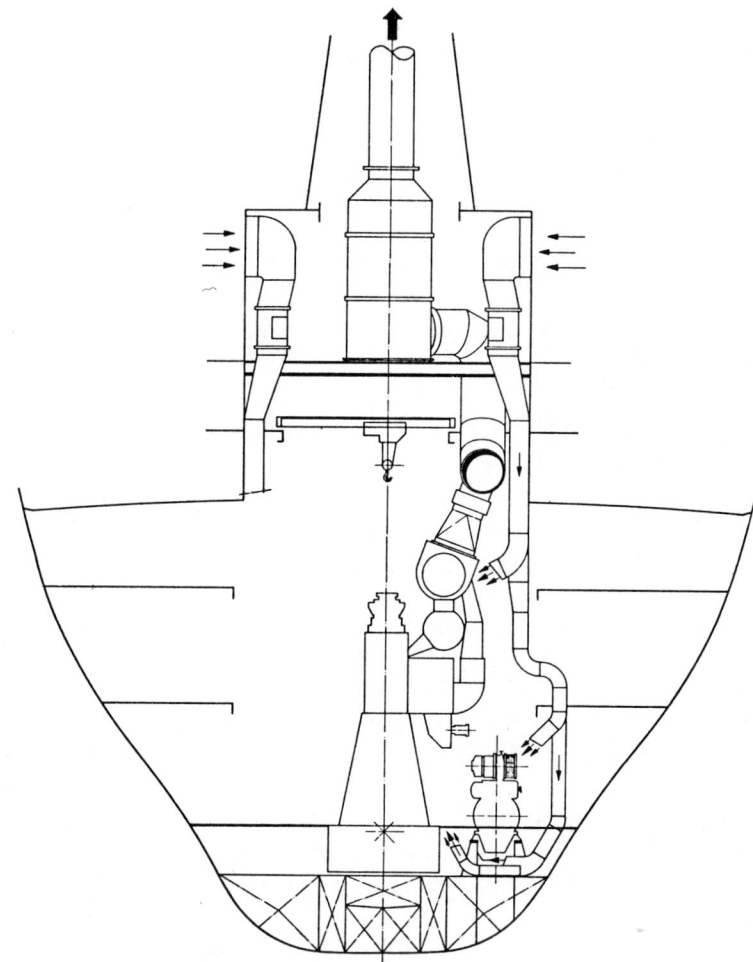

Figure 10.10 Arrangement of ventilation for Sulzer RND..M engines (Sulzer drawing).

nies high back pressure. The consequence of excessive exhaust temperature may be unacceptable thermal stresses within the engine; hence all engine builders specify their maximum allowable back pressure (pressure above atmospheric at the turbocharger outlet). Sulzer, for example, specifies a maximum of 300 mm water (0.03×10^5 Pa) for the RND engines.

Figure 10.11 (Sulzer dwg 4-107.095.215-2) shows the exhaust connec-

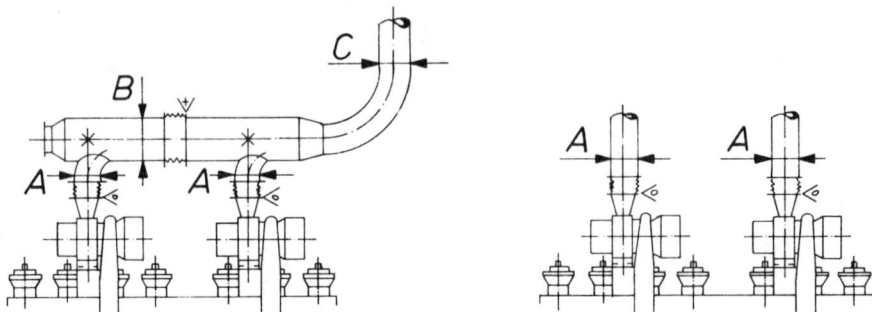

Figure 10.11 Arrangement of exhaust line for Sulzer RND..M engines (Sulzer drawing).

tions to the Sulzer RND90M engines. Note the locations of the expansion joints indicated.

10.8 EXHAUST GAS WASTE HEAT

Approximately 30 percent of the fuel energy input to a low speed engine is discharged into the exhaust system as thermal energy. Because the gas is much hotter than the surroundings (the sea, in particular), the potential is present for converting this energy into useful work in some nondiesel engine. For instance, a Carnot engine working between a possible gas temperature of 300 C and a 25 C seawater heat sink has an efficiency of about 48 percent, or 14.4 (0.48x30) of the fuel energy input. Since the diesel shaft output may be 41 percent of input, this estimate seems to show the potential for a 35 (14.4/0.41) percent increase in shaft output.

The estimate just made can be only an upper limit—not at all a realistic figure—for many considerations stand between it and actual use of the exhaust energy. Nonetheless, it shows that a large energy resource is there to be mined, and in practice it is often feasible to produce enough steam to run a turbine generator that handles all ship service electrical needs, while leaving sufficient steam for all direct heating services (recall that heavy fuel requires heating to reduce its viscosity).

The most prominent factors that reduce the ideal energy recovery are these:

> The gas is not the isothermal heat source required for the Carnot cycle, but cools as its heat is extracted.

EXHAUST GAS WASTE HEAT

The energy-using machinery cannot approach the Carnot level of efficiency.

Temperature differences between the exhaust gas as the using fluid (steam) reduce availability of the energy.

The engine seldom operates at its maximum continuous rating. The resulting lower gas temperature means a lesser availability.

Figure 10.12 is a plot of exhaust gas flows and temperatures for the Sulzer 6RND90M engine as a function of load (load = power along a cubic propeller curve), indicating the decline in gas quantity and temperature at lower loads, and incidentally, how ambient air temperatures influence these parameters.

The gas cools, of course, as heat is extracted, for it is not changing phase. Unfortunately, the cooling must be limited to avoid the condensation of acid-bearing moisture (sulfuric acid from sulfur in the fuel). The temperature leaving the waste heat exchanger ("waste heat boiler") is therefore usually not less than 180 C. Some loss in temperature must be expected between engine and the heat exchanger, say, 10 degrees. Reading 290 C from Figure 10.12 (85 percent load, 27 C ambient) and sub-

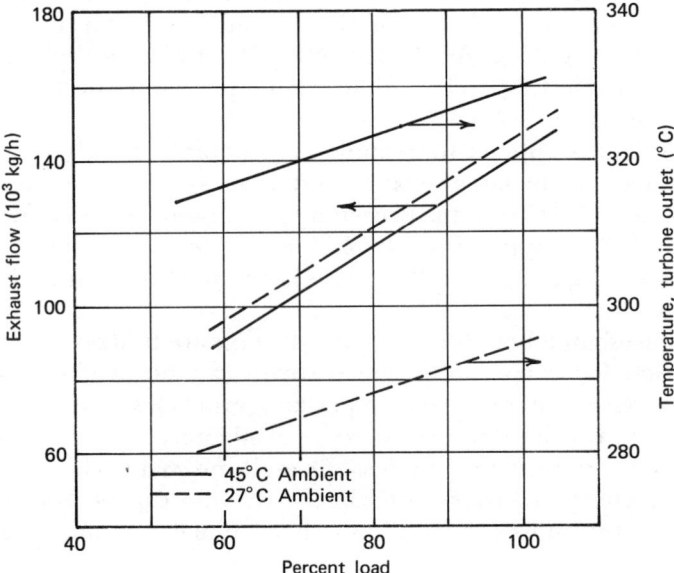

Figure 10.12 Exhaust gas flows and temperatures, Sulzer 6RND90M engine (from Sulzer data).

tracting this 10 degrees leaves a temperature drop of 100 C in the boiler. Using the flow for the same conditions (129,000 kg/h), and a specific heat of 0.96 kJ/kgC,* the energy extracted from the gas is 12.38x10⁶ kJ/h, which is approximately 11 percent of the fuel energy input (8770 kJ/kWh × 12,570 kW).

If this 11 percent is to be converted to mechanical or electrical energy (via mechanical means), only a small part of it is actually available. Using the 27 C ambient as sink temperature, the energy fraction that can be used in the limiting Carnot cycle is (temperatures in degrees K)

$$E = \frac{\int_{453}^{553} \frac{T - 300}{T} dt}{(553 - 453)} = 0.40$$

(= 40 percent of that 11 percent)

This calculation assumes a large number of small Carnot engines, each receiving heat at the gas temperature. As a practical matter, only the Rankine cycle can be used, and that not in large numbers—only one or two separate cycles in practice. The working fluid (water) for this cycle must evaporate at approximately the leaving gas temperature (see Figure 10.13). If it evaporates at 180 C after preheating from 27 C, approximately 4625 kg of steam can be produced by the engine at 85 percent propulsion load, and if the turbine generator that uses the steam is 60 percent efficient, about 490 kW of electrical power can be produced. This is 1.6 percent of the engine fuel input, or 3.9 percent of its shaft output.

If the steam is applied to heating, considerably more energy can be obtained because the latent heat is used. Let's say that steam for heating is produced at 180 C as in the preceding paragraph, and is condensed for heating at 100 C. Approximately 5250 kg of steam is produced, and its heat release is 12.38 × 10⁶ kJ/kg. This is about 11 percent of the engine fuel input.

The figures cited here for steam production are functions of the pressure chosen for evaporation; more steam can be produced at lower pressures, but its energy available per kilogram is less. Otherwise, some superheating can be done to increase availability, but the quantity of steam produced must then be less. Overall, no major change from the figures for energy recovery cited in this section can be gotten by juggling pressures and degrees of superheat, but if it is important to re-

---

*Specific heat is a function of air/fuel ratio and of temperature. Charts for its value can be found in textbooks on internal combustion engines, and in the Naval Architects and Marine Engineers diesel bulletin [Society of Naval Architects and Marine Engineers (1975)].

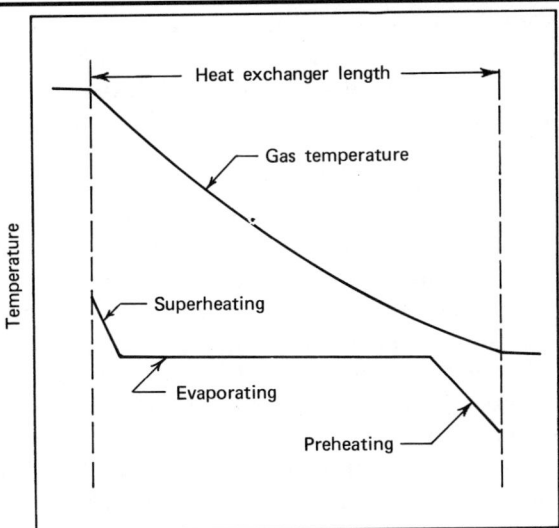

Figure 10.13 Typical temperature distribution in exhaust gas heat exchanger.

cover the maximum amount of exhaust energy, then a designer can optimize that recovery somewhat by judicious choice of pressure and temperature.

In the usual situation, steam is produced for both electrical power and for heating. High pressure and some superheat enhance efficient use of the steam by the turbine generator, while production of steam at the low pressure needed for heating maximizes the use of energy in this service. These conflicting requirements can be resolved in several ways. One is to produce all steam at conditions favorable for power generation, then use turbine exhaust, or steam extracted from the turbine, for heating. The other is to produce the steam at two different conditions—in effect, to have two separate heat exchangers in the exhaust gas stream, with the one operating at the higher pressure being the first contacted by the gas. On the other hand, if the supply of steam so exceeds demand that efficient use is not a consideration, then another alternative is to produce all steam at a high pressure for the benefit of the turbine, and to serve the heating needs from the same source via a reducing valve.

If steam is to be used at all, almost certainly a source independent of the engine will be essential, since the demand for steam is likely to continue when the propulsion engine is not operating. Ship service electrical power demand, for example, is only moderately reduced when the propulsion engine is not running. For this reason, supplementary production of steam by oil firing is essential, and is accomplished in several ways. One is to install separate boilers ("donkey boilers") that supply a common steam main, and are put on the line either manually or automatically when the pressure falls from a decline in the exhaust gas source. Another is to provide burners within the shell of the exhaust gas boiler, with these to operate as needed to supplement the exhaust source. A third alternative is to connect the water sides of the waste heat boiler and a separate oil-fired unit. A pump continuously circulates the water and water/steam mixture between the two units, and under automatic control of steam pressure, the burners of the oil-fired boiler contribute supplementary energy as necessary. Under this last scheme, separation of steam from the water takes place in the drum of the oil-fired boiler, and hence it is the source of all steam, whatever the origin of the heat. The exhaust unit is therefore often referred to as an "exhaust gas economizer."

In short, there are many alternatives in the choice of heat recovery equipment. Figure 10.14 is borrowed from a reference [Norris (1964)] that treats the alternatives in greater detail than is feasible here. The figure shows the simplest boiler (sketch (a)), a dual-pressure boiler (sketches (c), (d)), and several versions of the combined water circulation scheme (sketch (g), for example).

The steam circuits necessary to implement the waste heat idea can take on many configurations; although the basics such as condensers, condensate pumps and feed pumps may always be present, the details and degree of complexity vary widely. We shall not attempt to explore the many possibilities here, but only offer a sample. See Figure 10.15 [Norris (1977)]. Study of the diagram will show a dual-pressure system, but with only the high pressure steam produced directly by exhaust heat. This steam is superheated and drives a turbine generator in a simple circuit. Water is circulated from the drum (unit (6)) through economizer and evaporator sections, and along the way exchanges heat (in unit (1)) with the water of the "packaged boiler" (unit (13)). This boiler, unfired when the exhaust heat source is sufficient, produces low-pressure steam for heating services, and with the unit (1) heat exchanger therefore as the sole source of heat. Note that this illustrates yet another way of obtaining steam at two different pressures.

10.9 ELECTRIC LOAD FOR ENGINE AUXILIARIES

The motors that drive the engine auxiliary components obviously require electrical power. The amount depends on the size of the engine, for a larger engine must be served by greater flows of cooling water, lube oil, fuel, and ventilation. It is not feasible, however, to specify the electric load solely on the basis of engine data, since pumping power depends on

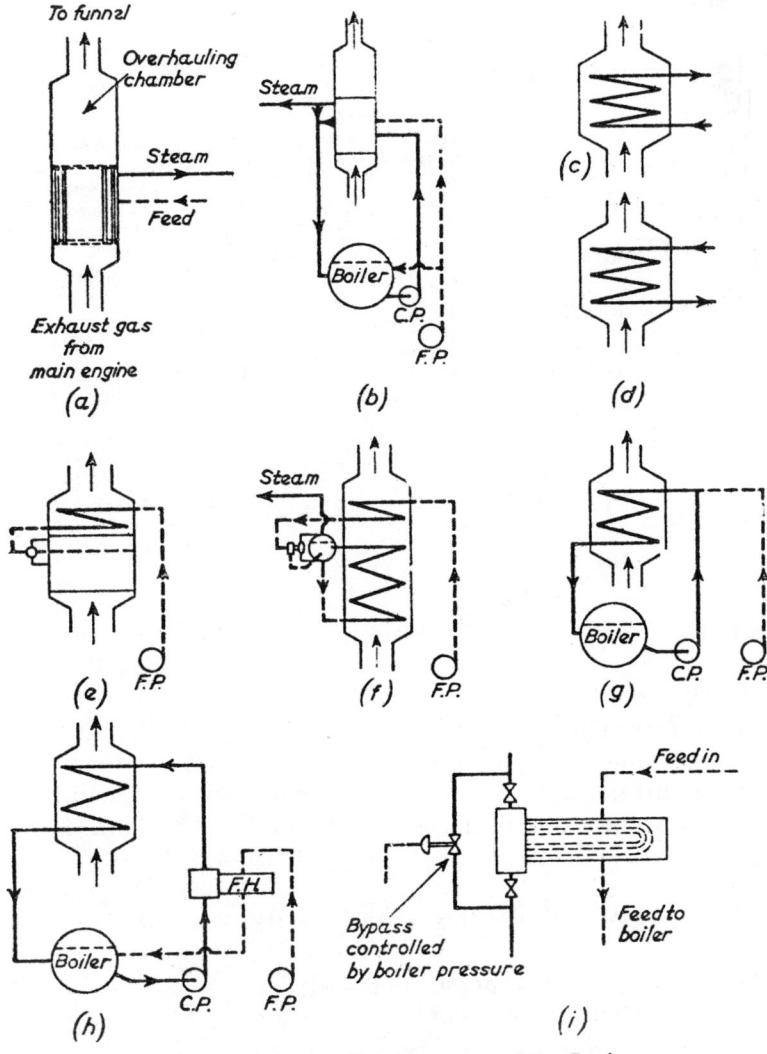

Figure 10.14 A selection of types of waste heat exchangers [from Norris (1964)].

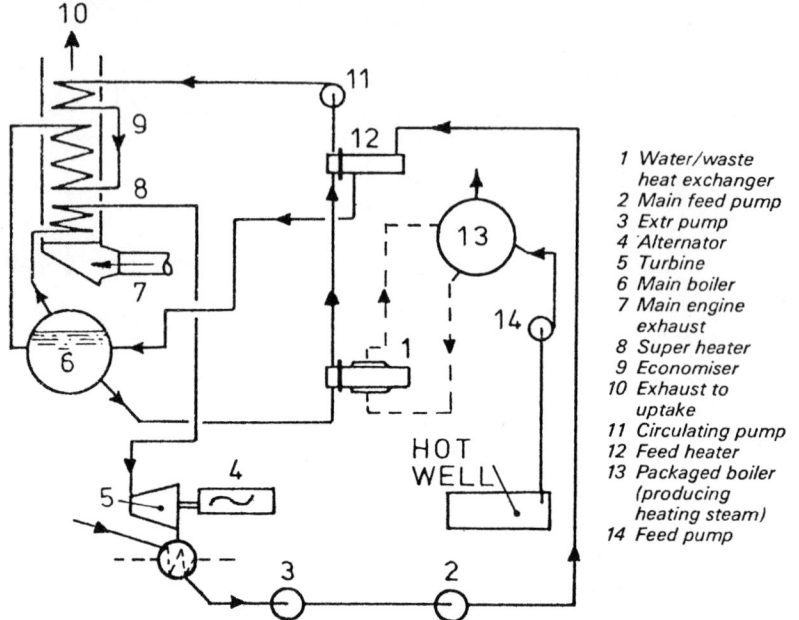

Figure 10.15 A simple two-pressure steam system [from Norris (1977)].

head as well as on flow, and head requirements depend on the configuration of the entire auxiliary system; identical engines in different ships may not have the same power demand. On the other hand, their demands must be roughly the same, and an estimating formula can be offered, that of the Naval Architects and Marine Engineers diesel bulletin [Society of Naval Architects and Marine Engineers (1975)], as follows:

$$kW = 0.008 \text{ shp} + 1.6N + 9\sqrt{N} + 80 \qquad (10.2)$$

where N = number of persons aboard
        shp = shaft horsepower

Note that this is intended to give the entire electric load for the ship.

An example of engine auxiliary power loads for a particular ship appears in Table 10.1. Note that this includes *only* engine-related loads.

10.10 MACHINERY WEIGHT

Weights of propulsion engines are generally published by their builders, and therefore are readily obtainable. Total machinery weights are less

TABLE 10.1 SUMMARY OF ELECTRICAL LOAD REQUIRED BY PROPULSION ENGINE AUXILIARIES[a]

Auxiliary	Load (kW)
Seawater circulating pump	138.8
Jacket water circulating pump	52.0
Piston water circulating pump	29.7
Fuel valve cooling pump	2.1
Fuel oil booster pump	7.0
Lube oil pumps	49.0
Turbocharger cooling pump	6.0
Fuel oil purifier	10.7
Lube oil purifier	4.6
Auxiliary air compressor	46.7
Control air dehumidifier	0.4
Sludge incinerator	20.0
TOTAL	367.0

Source: Levine et al. (1979).
[a](Sulzer 10RND90 engine, 29,000 bhp, 150,000 dwt tanker.)

apparent, since they depend upon information from designers and builders of the ships, and these people have little reason to publish such details. From a rather barren literature, then, two examples are offered here, Tables 10.2 and 10.3.

10.11 SHAFT DRIVE FOR AUXILIARY POWER

You will have noted that material in this chapter is limited to auxiliary items associated directly with the engine, but Figure 10.16 illustrates an exception: a power takeoff gear for the propeller shaft. Electric power generators driven by the main engine are fairly common in marine propulsion technology, for the fuel rate of that engine is usually better than that of a smaller auxiliary engine. On the other hand, this expedient is rare when the engine is a low speed diesel, since there is no reduction gear to effect the speed change (nearly always required) between engine and generator. The figure, however, shows a gear designed for power takeoff service with a low speed engine. It fits into the shaft line, in effect replacing a short piece of shaft, and produces a higher speed at its output

TABLE 10.2 SUMMARY OF PRINCIPAL MACHINERY WEIGHTS FOR PROPULSION PLANT

Machinery	Weight (tons)
Propulsion engine with thrust bearing	1130
Propulsion engine auxiliaries	100
Diesel generators and turbine generator	57
Exhaust gas boiler, oil-fired boiler and their auxiliary components	437
Cargo pumps (4, rated 5000 m³/hr each)	38
Propeller and shafting	103
Piping (including exhaust) associated with propulsion engine	145
Piping (including exhaust) associated with boilers	89
Gratings, ladders, workshop, ventilation	95
Spare parts	43
Liquids (not including fuel)	112
TOTAL	2349

Source: B & W study (unpublished).
[a]Burmeister & Wain 8K98FF engine, 32,000 bhp, 200,000 dwt tanker.)

coupling (on the back of the unit pictured, upper right). The unit in the figure, by Lohmann & Stolterfoht, increases speed from 122 rpm to 720 rpm, and is fitted with a pneumatic clutch for disconnecting the driven generator. The piping arrayed on the visible side of the gear set is part of its integral lubrication system.

10.12 FIRE PROTECTION

Ship machinery spaces are always fitted with fixed fire extinguishing systems, cylinders of carbon dioxide that can be triggered from remote locations in case of serious fire being the typical example. Such systems are not unique to the low speed diesel, and are not engine auxiliaries any more than, say, the bilge system is. However, we must note that some regulatory bodies require a similar system to be fitted directly to engine crankcases and scavenge air spaces. If required, Sulzer recommends eight 45 kg carbon dioxide flasks or three 90 kg Halon flasks for protection of the 6RND90M engine.

TABLE 10.3 SUMMARY OF PRINCIPAL MACHINERY WEIGHTS FOR PROPULSION PLANT

Machinery	Weight (tons)
Propulsion engine	621
Propeller and shafting	66
Lube oil system	19
Fresh water cooling system	10
Seawater cooling system	9
Fuel system	1
Fuel valve cooling system	2
Starting air system	22
Diesel generators	68
Exhaust systems	2
Waste heat boiler	18
Lubricating oil in tank[b]	82
TOTAL	920

Source: Neumann and Carr (1967).
Wain 84-VT2BF-180 engine, 14,400 bhp-, 55,000 dwt dry bulb carrier.
[b]All other liquids included in the respective systems.

Figure 10.16 Takeoff gear for low speed engines. To be mounted in propulsion line shaft (Lohmann & Stolterfoht photograph).

10.13 CLEANING SYSTEMS

Several parts of the engine may require frequent cleaning, and it is usually fitted with cleaning equipment that collectively might be classified as another auxiliary system. This equipment supplies water under adequate pressure for spraying through fixed or portable nozzles, and may include mixing tanks for cleaning agents to be added to the water.

Figures 10.16 and 10.17 illustrate the cleaning equipment fitted to Sulzer RND90M engines. The first of these serves the combustion air coolers, and the second serves the drain spaces around the piston telescopic cooling pipes and the piston underside area.

The functions of these units as part of engine regular maintenance are discussed in Chapter 11.

10.14 REFERENCES

American Bureau of Shipping (1980), *Rules for Building and Classing Steel Vessels.*

Femenia, Jose (1973), "Economic Comparison of Various Marine Power Plants," *Transactions,* Society of Naval Architects and Marine Engineers, Vol 81, pages 79-108.

Levine, R A, Sucharski, D B, Stewart, E V (1979), "Recent Operating Experience for Diesel and Steam Turbine Propelled Oil Tankers," Society of Naval Architects and Marine Engineers, San Diego Section.

Norris, Alan (1964), "Developments in Waste Heat Systems for Motor Tankers," *Transactions,* Institute of Marine Engineers, Vol 76, pages 397-429.

Norris, Alan (1977), "Waste Heat Recovery—Are We Too Conservative?" *Shipbuilding & Marine Engineering International,* Vol 100, pages 71-73.

Neumann, J and Carr, J (1967), "The Use of Medium-Speed Geared Diesel Engines for Ocean-Going Merchant Ship Propulsion," *Transactions,* Institute of Marine Engineers, Vol 79, pages 89-129.

Society of Naval architects and Marine Engineers (1975), *Marine Diesel Power Plant Performance Practices,* Technical and Research Bulletin 3-27.

10.15 NOTATION FOR CHAPTER 10

bhp	brake horsepower
C	degrees Celsius
C_p	specific heat (constant pressure)
dwt	deadweight tons
g/kWh	grams per kilowatt hour
kg	kilograms

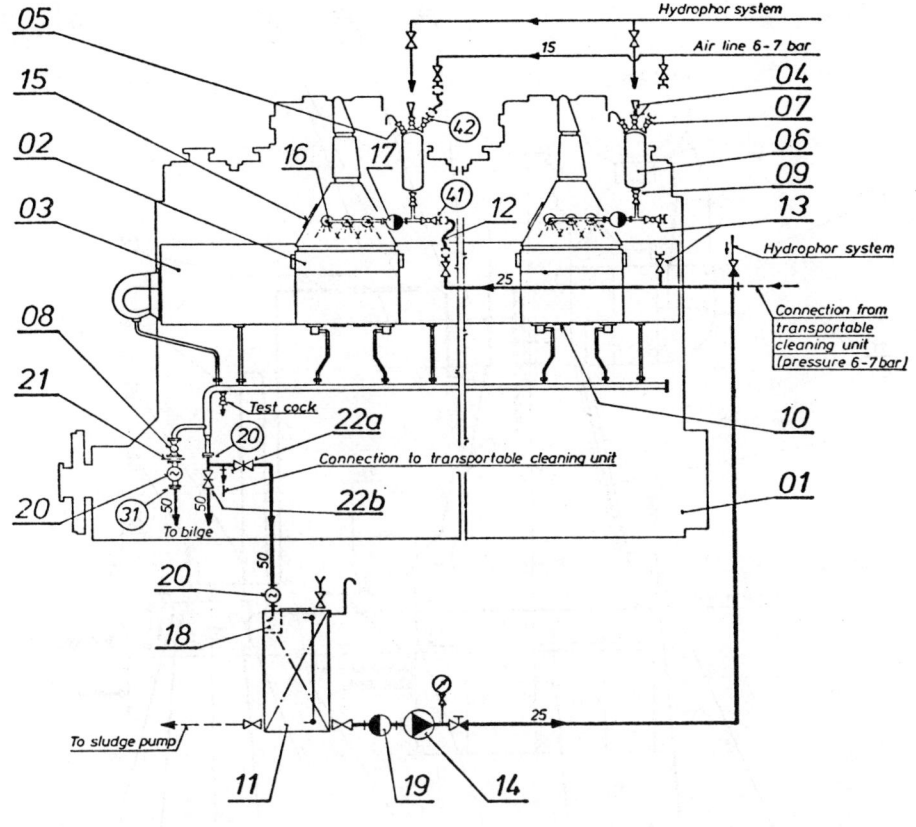

Figure 10.17 Cleaning system for air coolers (Sulzer drawing).

01 Main engine
02 Air cooler
03 Air receiver
04 Filling funnel and cock
05 Vent cock
06 Chemical dosage tank, capacity 26 l
07 Air supply cock
08 Drain cock
09 Discharge cock
10 Inspection cover on air receiver
11 Chemical tank, capacity approx. 500 l
12 Flexible hose
13 Discharge cock "in-port cleaning"
14 Circulating pump approx. 1.5 m³/hr × 6–7 bar
15 Inspection cover before air cooler
16 Nozzle
17 Filter before nozzles, mesh size 150 my m
18 Basket filter, mesh size 1 mm
19 Suction filter, mesh size 150 my m
20 Sight glass
21 Throttling disk
22a + b Sluice valve

211

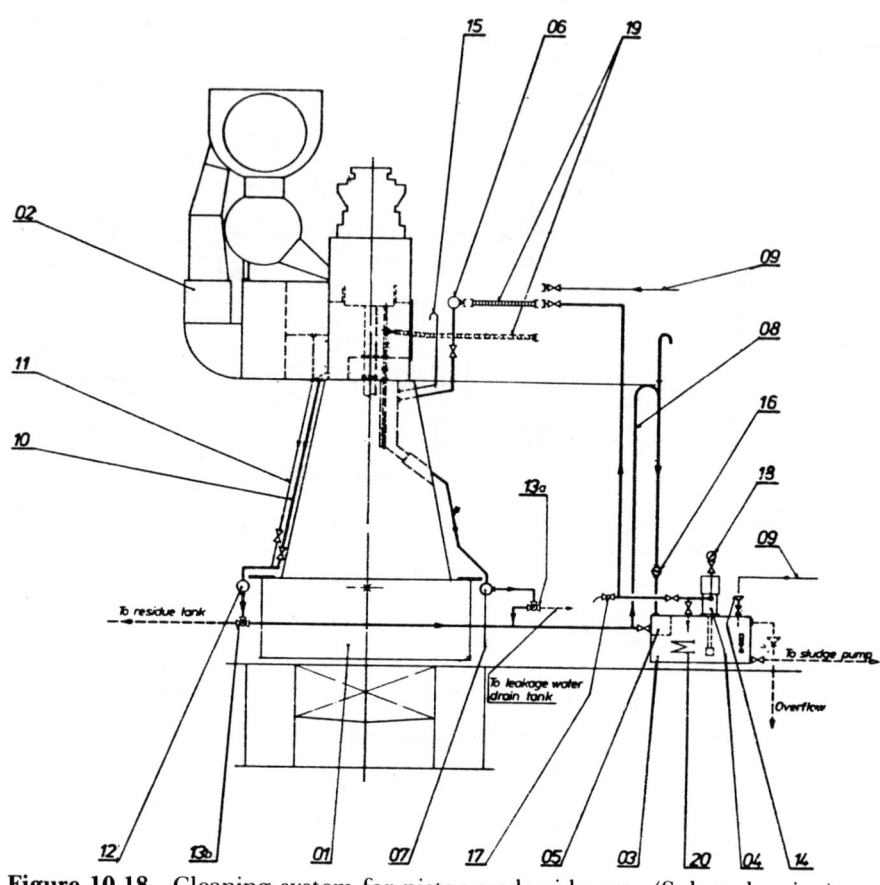

Figure 10.18 Cleaning system for piston underside area (Sulzer drawing).

01 Main engine
02 Air cooler
03 Hot water preparation tank
04 Hot water washing pump
05 Strainer
06 Hot water pipe for washing purposes
07 Leakage pipe of piston cooling leakage box
08 Water level setting pipe
09 Pipe from F.W. hydrophore system
10 Dirty oil from piston underside
11 Dirty oil from air receiver (for RND 76 + 90M only)
12 Common dirty oil outlet pipe
13a–b Three-way cock ("L" port)
14 Detergent inlet
15 Air vent from piston cooling leakage box
16 Sight glass
17 Test cock
18 Pressure gauge
19 Hose
20 Heating coil

NOTATION FOR CHAPTER 10

kg/h	kilograms per hour
kJ/kWh	kilojoules per kilowatt hour
kJ/kgC	kilojoules per kilogram degree Celsius
kW	kilowatts
$\dot{m}$	mass flow rate
m³	cubic meters
m³/h	cubic meters per hour
mm	millimeters
N	number of persons aboard
P	power
Pa	Pascals
rpm	revolutions per minute
shp	shaft horsepower
ΔdT	temperature difference

Chapter Eleven

ENGINE FAILURES, WEAR, AND MAINTENANCE

11.1 INTRODUCTION

The low speed diesel renders "good service," as attested by its great popularity over many years, all in competition with other machinery that also serves well. Nonetheless, high mechanical and thermal loadings, the many vagaries of service conditions, the general severity of marine service, and the multitude of moving parts all contribute to degradations of both gradual and abrupt character.

The gradual degradations are seemingly inevitable. Wear of moving mating parts, for example, cannot be eliminated, but only controlled by conservative design and adequate lubrication; if total wear reaches an unacceptable level during the life of an engine, the affected parts must be replaced. Periodic inspections and continuous condition monitoring, a ready supply of replacement parts, and provision of special tools that ease the tasks of disassembly and assembly are therefore recognized adjuncts of engine operation.

The abrupt degradations, or "failures," should not occur, but occasionally do—piston rings may break, for instance, or a crack may develop in a cylinder cover. Many such occurrences are discovered by regular inspection and monitoring, and replacement of the faulty part is made in orderly fashion when the engine is next available for maintenance work.

Or rarely (one would hope), a catastrophic failure occurs, a failure that stops the engine at sea, requiring emergency repairs by the engine crew, or (very rarely) a tow to the shipyard.

This chapter looks at what the common degradations are, the frequency at which they may occur, and the provisions made to minimize their impact.

11.2 ABRUPT DEGRADATIONS

An abrupt degradation, or "failure," is an occurrence that the engine designer does not expect, or that happens much sooner than the designer expects. The causes, in a general sense, are easy to imagine: the design may be inadequate, a defect of manufacturing may occur, or a severe overload may occur under some unusual service condition. A common story is the "teething troubles" of a new engine model, largely due to the design not being adequate in all of its many details. A pattern of failures leads, of course, to rectifications by the builder, followed by a decline or disappearance of the trouble.

The frequency at which failures occur thus fluctuates. It should tend to decline as technology matures, but the introduction of new engines tends to reverse the decline. (And perhaps the ineradicability of "human failure" will forever keep it from declining to zero.) In the 1960s the rapid increase in tanker size inspired a simultaneous movement among the diesel builders toward engines of greatly increased output. As the ships powered by these engines came into service there was a sharp increase in casualties at sea reported by diesel ships. Such, at least, is the implication of casualty data reported by a group of Japanese investigators [Tamaki et al (1978)]. Their reported frequency of casualties (casualties per vessel-year) climbs rapidly in the early 1970s, and they identify this frequency as being distinctly higher among the high-output engines. After 1973 the frequency declines, and one may surmise that the abrupt drop in new buildings after the petroleum embargo of 1973 led to a general maturing of engine populations. An additional factor may be the general slowing of ship operating speeds after the major increase in fuel prices of 1973 and later years.

So what troubles can happen? One could answer this by saying "almost anything," since all parts are subject to stresses of one or more kinds, and hence are at some risk of failure. And indeed a search of marine engineering literature will uncover reports of a wide range of failures. On the other hand, we'll limit the discussion to those things that happen with sufficient frequency (even if that frequency is small) to deserve categorization, as following:

1. *Rapid foulings.* Formation of carbon deposits ("trumpets") around fuel injection nozzles. Buildup of combustion products in exhaust ports, or exhaust valves (if used), in piston-ring grooves, and on turbocharger turbine blades. Deposits in scavenge chambers (particularly piston underside compression spaces) from lube oil leakage.
2. *Cracking of highly stressed parts.* Cylinder covers, cylinder liners, pistons— these due to combination of thermal and mechanical stresses. Structural parts such as bedplates, especially cracks in welds. Crankshafts.
3. *Breakages.* Piston rings are perhaps the components most likely to break. Gear teeth, as in the drive trains for camshafts. And any component that cracks will eventually fail completely if not replaced or repaired.. Although cracks mentioned in 2 are usually discovered before progressing this far, an occasional broken crankshaft—a serious casualty, indeed—has been reported.
4. *Rapid wear.* Any bearing or other contact surface, such as cam rollers, that receives inadequate lubrication, or that may be attacked by a contaminated lubricant. Piston rings and cylinder liners are special cases because sliding contact is only part of their burdens. For example, blow-by of exhaust gas may scour away the lubricant film, leading to rapid localized wear.
5. *Burning.* Most likely to affect piston surfaces exposed to cylinder gas; rapid (and usually localized) oxidation destroys metal surfaces. Rapid thinning of turbocharger walls under attack of the hot exhaust gas could also be noted in this category.

And how often do such things occur? That is difficult to answer because of the paucity of published data, but it is surely true that frequencies fluctuate for the reasons (among other reasons, perhaps) discussed at the beginning of this section. It is apparently common for a new engine model to experience its own particular troubles for a while, but after improvements by the builder, they may disappear. One of our references [Aue (1966)] discusses a history of this nature for the Sulzer RD engine. However, this source gives no data on frequencies of occurrence.

At the time this was written (1979), the only recent reports citing frequencies, found in our language, were two Marine Engineering Society in Japan papers [Tamaki et al (1978), Sakamoto (1978)]. Table 11.1 is taken from the first of these, and epitomizes the information from these sources that is suited to the discussion here. It results from 1029 vessel-

TABLE 11.1 FREQUENCY OF PROPULSION ENGINE STOPPAGE CASUALTIES

Failure Part	Frequency (no. of casualties/ vessel-year)	Hour of Stoppage (hr/ vessel-year)	Mean Hour of Stoppage (hr/no. of casualties)
Cylinder cover	7.80×10^{-2}	0.553	7.09
Cylinder liner	2.84	0.489	17.22
Piston (including ring)	2.63	0.223	8.46
Piston cooling system	1.52	0.132	8.70
Exhaust valve	8.92	0.196	2.20
Fuel oil valve	1.01	0.005	0.48
Driving gear	1.22	0.061	5.00
Fuel oil pump	3.65	0.110	3.00
Exhaust control valve	1.11	0.027	2.42
Supercharger	5.78	0.557	9.65
Lubricator	0.91	0.030	3.27
Maneuvering gear	0.91	0.030	3.25
Piping and tank (including cooler)	1.52	0.085	5.59
Others	3.04	0.120	3.94
TOTAL	42.86×10^{-2}	2.619	6.17

Source: Tamaki et al (1978).

years of operation over a 16 month period in 1974 and 1975. A total of 694 casualties are included in the data.* The table shows the frequencies of casualties (casualties per vessel-year), and data on the resulting engine down times. You will note that the actual defects are not specified; is is only the part involved that is listed. Since any part may be liable to several —or many— kinds of faults, reporting in detail would likely be too cumbersome. Nonetheless, the table gives a fair idea of troubles that occur, or were occurring during the 1974-1975 sampling period.

One category in Table 11.1 ("piping and tank") apparently covers auxiliary components, but observe that this also entailed some stoppage time. An encouraging point is the absence of any data on crankshafts, this said because a failed crankshaft is probably the most serious failure that the engine can experience.

Let's close this section by taking note of the failures of a truly hazardous nature that can occur, that is, events that threaten the safety of the engine crew, and perhaps of the ship itself. Only two such events seem to have any history of occurrence, they being (1) fire and (2) crankcase explosion.

Fire external to the engine, involving fuel or lube oil, is a possibility because of the presence of large quantities of these combustible liquids. For this reason remotely operated fire extinguishing systems (such as a carbon dioxide flooding system, typically) are required by regulatory bodies, but the hazard is approximately the same for all types of machinery—there is nothing unique about the engine discussed here in this respect. Fires that originate in the engine itself are likely to occur (*if* they occur) in the scavenging air spaces at places where oil leakage can accumulate. For example, engines that use the piston undersides as compressors run the risk of cylinder lube oil leakage into those spaces. Some regulatory bodies require built-in fire extinguishing equipment to reduce the hazard. Good maintenance and "housekeeping" are efficacious preventives.

Explosion of oil vapor in the crankcase, ignited perhaps by a hot bearing, is possible. For this reason, spring-loaded explosion-relief valves are fitted, typically one per cylinder. Their action may prevent damage to the engine if an explosion occurs. Some regulatory bodies require built-in fire extinguishing equipment for the crankcase also, though such equipment is not likely to prevent an explosion that occurs before a fire.

---

*Apparently nearly all of the engines involved are low speed, though we are not given the exact percentage.

11.3 EXPECTED MAINTENANCE

Quite without unexpected degradations, many wearing parts have lives shorter than the intended life of the engine. They must be inspected periodically (or monitored continuously—recall Chapter 9), and replaced at the appropriate state of wear. And some parts, such as the pistons, are not worn in the usual sense, but suffer gradual degradations that demand replacement.

The major component that wears in an expected way is the cylinder liner. Indeed, the term "engine wear rate" is virtually synonymous with liner wear rate. That rate will obviously vary among engines, for it is a function of a number of factors, but a rate of 0.1 mm/1000 hr is widely regarded as an acceptable figure [Golgothan (1978)], and serves almost as a standard. The authority just cited states that the resulting liner life should thus be 7 years; on the assumption that this is about 50,000 hr of operation, total acceptable wear is about 5 mm. With such an increase in diameter, the piston rings begin to lose sealing effectiveness.

Piston rings also wear, of course, with the top ring usually having the greatest wear rate, and a rate somewhat greater than that of the cylinder liner. For example, data published by Sulzer [*The Motor Ship* (1978)] for its RND68M, RND76M, and RND90M engines show top ring rates (after about 5000 running hrours) in the range 0.1 to 0.5 mm/1000 hr, and liner wear rates in the range 0.02 to 0.05 mm/1000 hr.

The pistons themselves suffer wear only in the ring grooves, though burning away of metal on the severely heated top face can occur. Although a piston can be partially inspected through the cylinder liner ports, a complete inspection requires removal ("pulling") of the piston. This might be done as often as once a year, and such an inspection typically involves replacement of one or more piston rings. If it has suffered no unusual degradations, the piston will be replaced only once or twice during the life of the engine. Often a replaced piston is recycled by reconditioning of its worn ring grooves, and perhaps by replacement of its crown if burning has occurred.

Fuel nozzles are eroded by the extreme velocity of the fuel jets, and must be replaced before this erosion degrades the fuel spray pattern. A replacemnt cycle of 10,000 to 20,000 hr was typical in the 1960s [Aue (1966)]. Fuel pumps are also subject to wear in several of their parts, and replacement several times during the life of the engine is likely.

All bearings wear, but at a very low rate if satisfactory lubrication is maintained. Periodic inspection is required, not only by "good practice," but by regulatory bodies. Although many bearings will last the life of the engine, replacement of any bearing at least once during this life has to

TABLE 11.2 SCHEDULED MAINTENANCE ACTIVITIES, GMT 908S ENGINE

Item	Action	Frequency	Man-Hours per Engine
1. Cylinder covers	Chemically wash; grind seats to liners and valves	28,000	40
2. Cylinder liners	Clean, round off port edges	7,000	40
piston	Withdraw for general maintenance; renew rings	7,000	192
3. Piston rod scraper ring box	Dismantle; adjust scraper sectors; renew parts	7,000	32
4. Connecting rod crosshead bearings	Overhaul; adjust clearance; possibly replace	7,000	86
5. Connecting rod big end bearings	Overhaul; adjust clearance; possibly replace	14,000	64
6. Journal bearings	Inspect; adjust clearance	7,000	30
7. Thrust bearing	Inspect; adjust clearance	7,000	5
8. Linkage, crankshaft, guide, and shoe	Check deflections and clearances; possibly replace	7,000	24
9. Fuel pump case, cams, levers, and rollers	Generally inspect	7,000	32
10. Fuel pumps	Generally overhaul	14,000	32
11. Fuel valves	Replace	1,000	56

12. Fuel valve needle, guide	Clean; eventually replace	1,000	112
13. Fuel valve sprayer	Renew if worn	3,500	Included in 12
14. Starting valves	Generally overhaul	7,000	48
15. Safety valves	Generally overhaul; adjust pressure	7,000	400
16. Piston cooling elbow pipes	Adjust clearance; align	14,000	16
17. Heat exchangers	Clean; generally maintain	3,500	130
18. Scavenging air valve	Overhaul; clean	7,000	140
19. Air pump, stuffing box, and valves	Generally inspect	14,000	80
20. Starting air unit, governor	Generally inspect	7,000	24
21. Lubricating boxes	Routinely overhaul	7,000	64
22. Fuel and lube filters	Clean	7,000	24

Source: Cotti (1966).

TABLE 11.3 REPAIR AND MAINTENANCE ACTIVITIES, GMT 908S ENGINE

Item	Action	Frequency	Man-Hours per Engine
1. Cylinder covers	Generally overhaul and repair	10–12	a
2. Cylinder liners	Withdraw and repair for wear	7–8	64
3. Piston crown	Dismantle from piston and renew	3–4	80
4. Piston skirt	Dismantle from piston and renew	7–8	a
5. Connecting rod crosshead bearings	Re-babitt bottom half	2–3	a
6. Connecting rod big end bearings	Re-babitt upper half	5–6	a
7. Journal and thrust bearings	Re-babitt bottom half or one face of thrust	10–12	a
8. Fuel pump cams, rollers, levers	Adjust or change cams, rollers, bushings	10–12	20
10. Piston cooling elbow pipes	Renew bushings	4–6	32
11. Air cooler	Recondition tube bundle and covers	10–12	36
12. Oil cooler	Recondition tube bundle and covers	10–12	48

222

13. Water cooler	Recondition tube bundle and covers	10–12	36
14. Fuel valve water cooler	Recondition tube bundle and covers	10–12	2
15. Turbocharger	Renew ball bearings, oil pump	2	15
	Renew inlet and outlet exhaust gas cases	7–8	9
	Renew rotor and distributor	10–12	9
16. Zinc anodes	Replace	1	80
17. Main tie rods	Tighten	2	20

Source: Cotti (1966).
[a]Man-hours included in the Table 11.2 actions for these items.

be expected. Ball bearings used in the turbocharger are likely to require replacement more frequently.

Other than these replacements for wear, the major expected maintenance action is cleaning. Oil leaking from cylinders into scavenge air spaces is a fire hazard, and hence must be removed. The scavenge air components—turbocharger compressor and air cooler—are fouled by the mixture of oil vapor and dust that is likely to be present in machinery space air, and this fouling must be removed lest compressor performance and air cooler effectiveness be seriously degraded. And the turbocharger turbine is fouled by engine combustion products to the detriment of its performance unless cleaning is performed frequently.

Most engines are fitted with installed cleaning equipment that allows cleaning to proceed even while the engine is operating. For example, Chapter 10 has described the cleaning system fitted to Sulzer RND..M engines, a system that protects turbine, compressor, air cooler, and piston underside air spaces. Both turbine and compressor are cleaned by water spray, the dissolving and scouring action of the water being sufficient for the job. The air cooler spray water includes a cleaning fluid. All of these sprays are used with the engine running, and are used frequently (once a week, perhaps) since even slight fouling can degrade performances significantly.

The Sulzer cleaning system also includes an optional outfit for washing the air cooler when the engine is not in operation.

The air spaces under the pistons can be cleaned only when the engine is not in operation. The Sulzer system includes pumps and mixing tanks for the cleaning solution. Cleaning is done by a spray from hand-held hoses.

Now, the topic has been covered in a general way by the preceding discussions, but additional insight can be gotten from a summary of the expected maintenance activities. Although these activities and their frequencies differ somewhat among different engines, and have changed somewhat as the engines have evolved, a good illumination can be had by looking at the expected requirements for one engine at a particular time in its history. Tables 11.2 and 11.3 do just this; they are condensations of maintenance schedules published for GMT engines in the 1960s [Cotti (1966)] (engines of 900 mm bore, eight cylinders). Table 11.2 is a list of scheduled maintenance activities, with their scheduled frequencies when operating 7000 hr/year, and the expected man-hours. Table 11.3 is a list of overhaul work that may be necessary as the result of expected degradations discovered during the scheduled activities. Predicted frequencies of occurrence, and the resulting manhours, are also tabulated. These tables provide more detail than has been given in the previous discussion, though they are generally consistent with that discussion. The most noticeable discrepancy is in the fuel nozzle life (fuel *valve* in the tables), it being listed in the table as shorter than the life mentioned above. Different engines, different times—you will note that the source dates considerably before this publication. Again, though, the picture given by these tables is generally valid.

11.4 MAINTENANCE EQUIPMENT

Maintenance of a complex mechanical device inevitably requires an outfit of special tools. The low speed marine diesel is certainly no exception, and also has elements of uniqueness due to its size and due to its frequent use in ships that offer only short in-port intervals for major maintenance activity. For the engine crew to replace a piston, say, that may weigh several thousand kilograms, and to replace it without delaying a short-turnaround ship in port, requires an array of equipment to facilitate rapid working with massive parts. The equipment can generally be categorized into lifting items and those used for prestressing the many bolts and studs.

The first essential among lifting devices is the engine room crane,

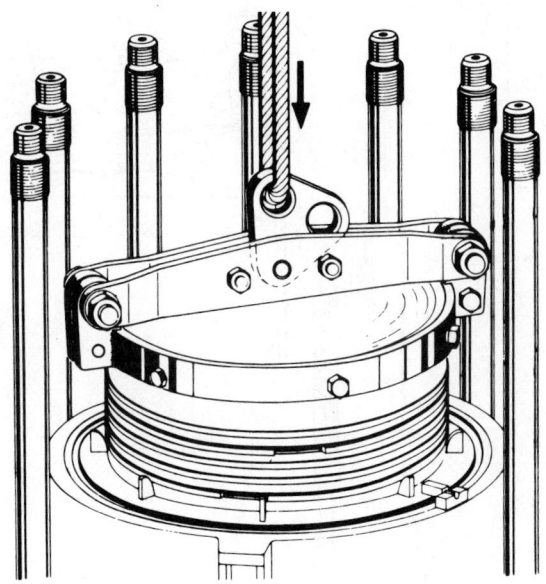

Figure 11.1 MAN piston lifting device (MAN drawing).

226   ENGINE FAILURES, WEAR, AND MAINTENANCE

always a bridge crane that travels on rails above the engine. Obviously its travel must encompass all parts of the engine and surrounding working areas, and its lifting capacity must be sufficient for the largest component to be handled. Cylinder cover, cylinder liner, and piston are the heavy components; all are of about the same magnitude in weight, with the cover usually being the heaviest. For example, the Sulzer RND90M cylinder cover weighs 5465 kg, and Sulzer recommends a minimum crane capacity of 6000 kg for this engine.

Special lifting fixtures may be necessary for the item being lifted. Figure 11.1 is an example, showing a fixture attached to the position of a MAN engine being lifted from its cylinder. After the cylinder cover is removed (seven of the cover studs are quite evident in the picture), and the piston brought to top center, the band shown fitted around the piston can be fastened to it by machine screws.

Figure 11.2 shows another piston lifting device in action. It is a Sulzer

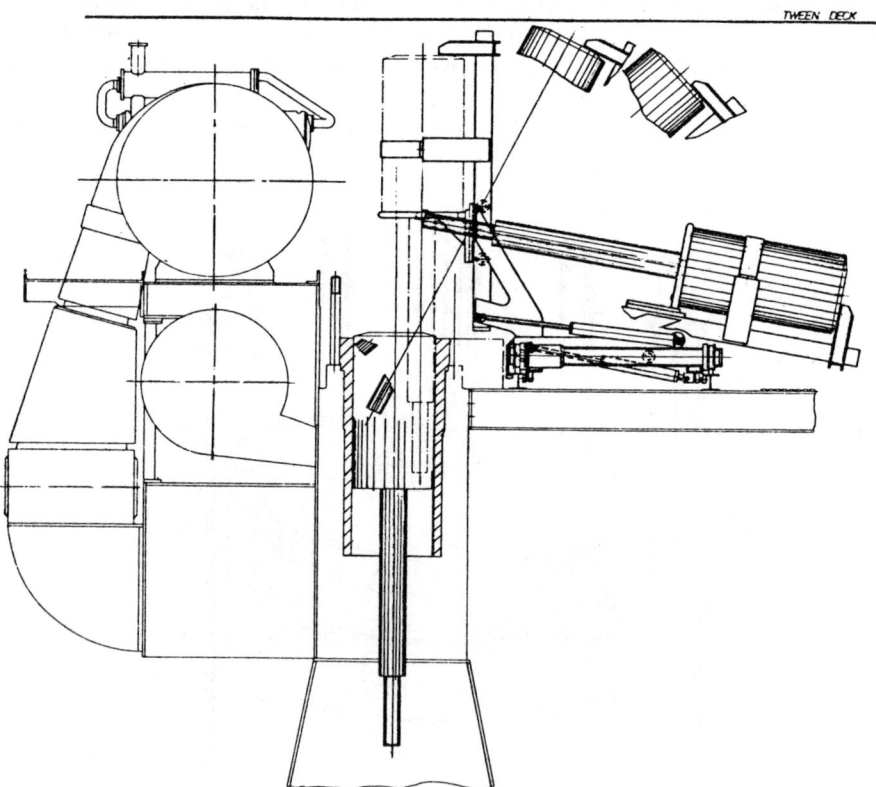

Figure 11.2 Piston removal fixture for Sulzer engines (Sulzer drawing).

MAINTENANCE EQUIPMENT 227

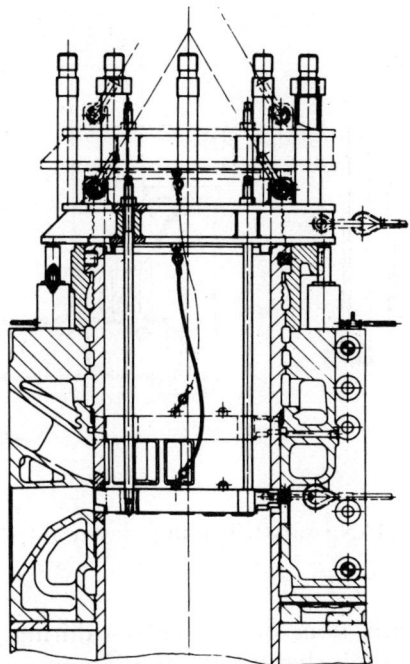

Figure 11.3 MAN cylinder liner removal device (MAN drawing).

product, and is used in conjunction with a special crane (also evident in the figure) when overhead clearance is too small for the usual straight lift of piston and piston rod.

Figure 11.3 shows a lifting fixture for MAN cylinder liners. This device consists of a beam set across the top of the liner, a lower beam fitted through the scavenge ports, and tie rods clamping the two beams together.* Two hydraulic jacks sit on the cylinder jacket and press upward on the ends of the upper beam, this to overcome the fit between liner and jacket.

A lifting device of quite another type and purpose is pictured by Figure 11.4, it being a device to lift the crankshaft of MAN engines—not, by any means, to remove the crankshaft from the engine, but to lift it slightly to facilitate removal of a bearing shell. Bearing shells are normally removed by "rolling out," a process that consists of sliding the shell around to the top of the journal after the bearing cap has been removed. If the bearing

---

*Both of these beams appear twice in the figure, since it shows two stages of the lift—sort of a double-exposure view.

228   ENGINE FAILURES, WEAR, AND MAINTENANCE

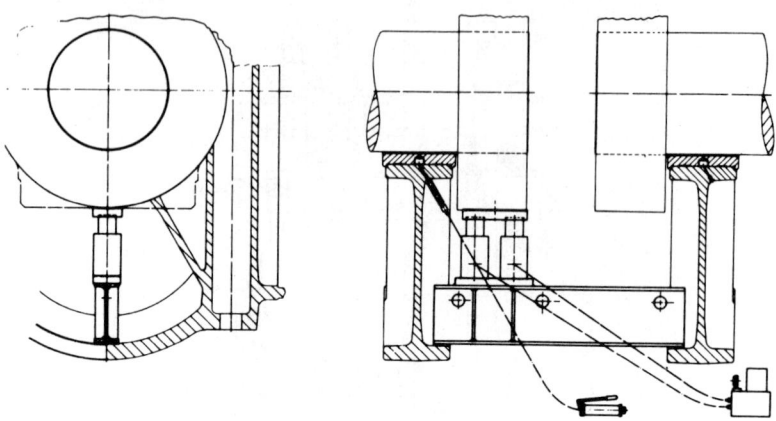

Figure 11.4 MAN crankshaft lifting device (MAN drawing).

seizes—as it apparently does occasionally— during removal, the only recourse is to lift the shaft slightly (0.2 to 0.3 mm) to free it. The figure shows the lifting fixture consisting of an I-beam placed across the bottom flanges of two bearing pedestals, supporting two hydraulic jacks that press upward on a crank web. The figure also includes a grease gun attached to a pocket under the bearing shell, this for the purpose of giving it a hydraulic lift to break the seizure.

All threaded connectors (bolts, studs, machine screws) should be given the correct pretensioning during assembly; for those that secure parts in proper alignment against the dynamic loadings that permeate diesel structure, the correct pretensioning is vital.

For smaller fasteners, an accurately calibrated torque wrench applied to the nut is sufficient. However, the traditional torque wrench has been replaced at many points by hydraulic nut tighteners. Figure 11.5 shows such a device set up to tighten connecting rod bolt nuts on a Sulzer RND..M engine. The hand-powered hydraulic pump appears in the foreground. Compared to the mechanical wrench, this device is handier to use in a limited space, is faster since it is applied to several nuts simultaneously, and for the same reason produces more nearly uniform tensioning among the several bolts.

A more elaborate hydraulic nut tightener is shown in Figure 11.6. The device shown applies simultaneous torque to the 16-cylinder cover stud

MAINTENANCE EQUIPMENT 229

Figure 11.5 Hydraulic tightening of Sulzer connecting rod bolts (Sulzer photograph).

nuts of Sulzer RND..M engines. Its use greatly speeds the removal and replacement of cylinder covers (and hence speeds any work involving access to the cylinder), as well as giving the benefits of uniform tensioning.

Correct tensioning can also be obtained by conventional wrenches with direct measurement of fastener elongation. Where this is intended, special measuring fixtures must be provided. An alternative is to elongate by heating, with the nut being turned, virtually torque-free, to the final position while the stud or bolt is at high temperature. When this is intended, the fastener must have a hollow bore for the insertion of an electrical heater, and temperature and required angle of nut rotation must be known. Figure 11.7 shows this process being applied to connecting rod bolts of a MAN engine (cf Figure 11.5, which shows the same job being done hydraulically on a Sulzer engine). The heater inserted into the center of the bolt shows in the right-hand sketch. The left-hand sketch is a cross section showing the nut-turning handles applied to nuts on

230 ENGINE FAILURES, WEAR, AND MAINTENANCE

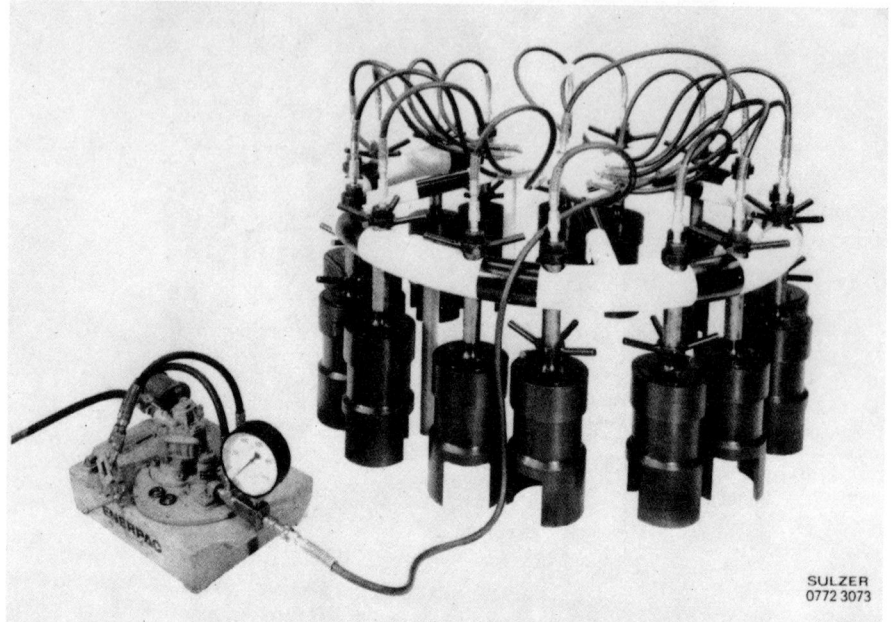

Figure 11.6 Hydraulic nut tightener for cylinder covers of Sulzer RND..M engines (Sulzer photograph).

opposite sides of the rod. Note that they are designed to rotate through exactly 90 degrees, that being the angle required to give the desired bolt tension after that member cools to normal temperature.

11.5 SPARE PARTS

Both scheduled maintenance and the response to failures usually require spare parts. The builders of low speed engines maintain extensive inventories of these necessaries, as do shipyards and repair firms, and given air transportation, a new part is never far away. Nonetheless, a stock of spare parts on board is a likely saver in time and transportation costs, and indeed is required by regulatory bodies. The required or desired list, when all parts are shown, is quite extensive, and we'll not attempt such a list here (after all, in its details it will not be the same for every engine anyhow). Table 11.4, however, will give you an idea of the extent of parts to be carried, since it includes the major items. It is a condensation of the lists recommended by several manufacturers, and equals or exceeds regulatory body requirements.

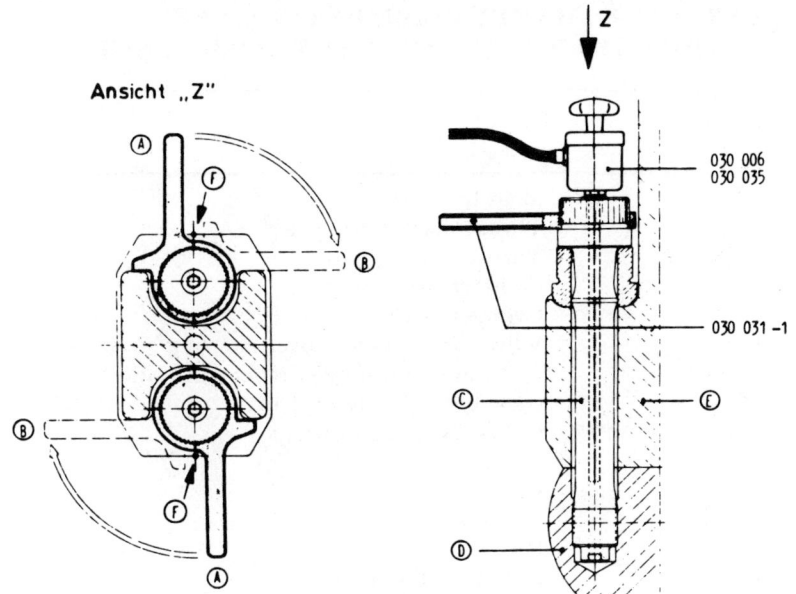

Figure 11.7 Thermal elongation of MAN connecting rod bolts (MAN drawing).

11.6 MAJOR REPAIRS *IN SITU*

A worn or damaged piston, cylinder liner, bearing, etc, can be removed from the engine and ship, and replaced readily by a new or reconditioned part. Not so the crankshaft. This large and weighty object (150 tons, perhaps) is expected to last the life of the engine, for its replacement requires dismantling of the engine, and is likely to require cutting of a hole in the hull for access. Fortunately, some of the unexpected degradations that do occasionally occur can be repaired without removing the crankshaft—so-called *"in situ"* repairs.

These repairs typically take the form of machining of crankpins or journals, perhaps (for example) in response to scoring or corrosion from a contaminated lube oil. Awkward though the task may be, portable machine tools can be rigged within the crankcase to perform the machining.

Such jobs are not likely to be done by the engine crew, nor by equipment furnished with the engine. Rather, this type of work is the province of specialized repair service companies that dispatch machinists and machines to the repair site from a central depot. In the 1970s several companies of this nature were operating from European headquarters.

TABLE 11.4 MAJOR SPARE PARTS TO BE CARRIED ON BOARD FOR A LOW SPEED ENGINE

Number	Part
2	Main bearing shells
1	Thrust shaft bearing shell
1 set	Thrust bearing pads
1	Cylinder liner
1	Cylinder cover
1 set	Valves for cylinder cover (starting air valve, fuel valve with nozzle, relief valve, indicator cock, blow-through valve)
1 set	Connecting rod bearings
1	Piston
1	Piston crown
1	Piston skirt
1 set	Piston rings
1 set	Piston cooling pipes
1 set	Camshaft driving gears
1 set	Camshaft bearings
1	Cylinder lubricator
1	Fuel pump
1 set	High-pressure compound fuel pipes
1 set	Exhaust system expansion joints

11.7 REFERENCES

Aue, George K (1966), "Operating Experience with Large Sulzer Diesel Engines," American Society of Mechanical Engineers, Paper 66-DGEP-15.

Cotti, Ernesto (1966), "Recent Experience in the Large Marine Diesel Engine Maintenance," American Society of Mechanical Engineers, Paper 66-DGEP-13.

Golothan, D W (1978), "A Review of the Causes of Cylinder Wear in Marine Diesel Engines," *Transactions,* Institute of Marine Engineers, Vol 90, Series A, pages 137-163.

The Motor Ship journal (1978), Vol 59, No 700, page 103.

Sakamoto, Yukio (1978), "Recent Tendency of Machinery Troubles," *Bulletin* of the Marine Engineers Society in Japan, Vol 6, No 3, pages 179-184.

Tamaki, H, Kurosu, K, Kagoshima, N, Iijima, H (1978), "An Analysis on Casualties in Ship Operations," *Bulletin* of the Marine Engineers Society in Japan, Vol 6 No 4, pages 324-337.

11.8 NOTATION FOR CHAPTER 11

GMT   Grandi Motori Trieste
MAN   Maschinenfabrik Augsburg Nurnburg
mm   millimeter(s)

Chapter

Twelve

THE ENGINE AND ITS ENVIRONMENT

12.1 INTRODUCTION

The propulsion engine's environment has three layers: the close-by technical one of structure and machinery to which it is intimately connected, the slightly more distant one of shipboard living and working spaces, and then the surrounding sea and atmosphere.

The first of these is the most important from the standpoint of successful ship operation, and is the one that receives the most attention here and in other chapters. It involves the interaction of the engine with its propeller and many auxiliary systems, a topic that has been covered by other chapters. The arrangement of the machinery is a major concern, and includes both the propulsion engine, its auxiliaries, and nonpropulsive machinery. This topic is covered here, mainly through graphic examples. Also in this category is the engine "seating," or its connection to ship structure, and it is likewise treated in this chapter.

The most nettlesome environmental problem of the first category is the one of vibrational excitation by the engine. The continuously varying torque of each cylinder and the rotating and reciprocating masses provide excitation to the shaft and to the ship structure. Vibrational analysis is consequently an essential part of propulsion plant design, but unfortunately is such a large topic that this chapter can give only a sketchy treatment. On the other hand, references for the omitted details are readily available.

The environment of shipboard living and working spaces is principally affected by the noise of the engine. Although some measures to reduce the strength of this source are possible, the major protection from engine noise must come from acoustical treatment close to the area to be protected. The subject of shipboard acoustical design is well beyond the scope of a book such as this, but the situation with respect to the engine and and surrounding spaces is summarized here.

Now, the more distant environment of sea and air receives only minimal impact from a diesel propulsion engine. There are no discharges to the sea, and by present (1980) standards, the exhaust gas discharge has no degrading impact on the atmosphere, and receives notice only when a ship makes excessive smoke in port. However, because of the environmental concern that a large power source generates, the topic is given a brief treatment here.

12.2 ENGINE ROOM ARRANGEMENT

Merchant ship propulsion machinery is usually located far aft in the hull, or in a position intermediate between amidships and the aft extremity. The relative merits of different positions depend on several factors of ship design that are beyond the scope of our discussions; we note only that the low speed engine is generally as suitable as any other type for a desired fore-and-aft location. Its most likely handicap with respect to location comes from its height. If the ship requires continuous decks low in the hull, as is usual in a roll on - roll off vessel, the height of a low speed engine may be difficult to accommodate, and a low-profile medium speed engine may be chosen instead.*

The engine sits upon the engine room tank top (inner bottom) on the centerline (there can be exceptions to the latter, of course, such as twin screw arrangements). The height of its output coupling is usually suitable for the propulsion shaft line without need for much rake of the shaft or for elevation of the engine via a foundation. The location of the engine within the machinery space is therefore fixed without much choice being available to the ship designer.

Because of its size, the propulsion engine dominates the engine room

---

*In the late 1970s, several engine builders were attempting to reduce the effective heights of their low speed engines to make them more suitable for roll on - roll off ships. For example, Sulzer introduced a crane and handling fixtures that require less overhead space for removal of pistons.

and effectively determines its dimensions. Other equipment, although numerous, can usually be fitted around the main engine with little difficulty for available space. The only common major exception to this statement is the exhaust system. The combined bulk of ducting, waste heat boiler, and spark arrester is comparable to that of the engine itself, and therefore requires a large volume above the engine.

"Overhaul space"—the space left empty so that parts can be removed without collision with structure or other equipment—is a requirement for all types of machinery. A diesel engine requires this space particularly for the removal of pistons and cylinder liners; hence there is always a large empty space directly above the engine. Chapter 11 has noted the need for an overhead crane for the handling of such parts. This crane must therefore be mounted in correct juxtaposition to the cylinders, and with sufficient clearance for its lifts.

Several figures should give you a good idea of the general arrangements of machinery in a typical ship, and show where many of the individual items might be placed. These figures, 12.1 through 12.4, are based on a proposed arrangement for a 36,000 ton tanker, powered by a Sulzer 6RND76M engine, and under construction as this chapter was being written. Several views are shown, and they are repeated to emphasize the several prominent auxiliary systems that are discussed in Chapter 10.

The great majority of ships powered with low speed engines are single screw, and Figures 12.1 through 12.4 are quite representative of them. But there are exceptions: some twin screw ships, and even a few triple screw, have been built since 1970. Figures 12.5 and 12.6 are included to show you how the engines might be fitted into the respective engine rooms in such cases.

12.3 ENGINE MOUNTING IN SHIP

Ship machinery is usually mounted on a foundation raised from deck or inner bottom. In the case of the low speed engine, however, the bottom structure itself forms the foundation—the engine sits upon the inner bottom. The designer of the ship structure must obviously provide bottom plating and stiffeners adequate in strength to support the weight of the engine, but must provide for several other factors, namely:

1. Propeller thrust reaction is transmitted to the hull via the engine thrust bearing, hence via the engine foundation.
2. Shaft torque reaction must also be sustained by the engine and its foundation.

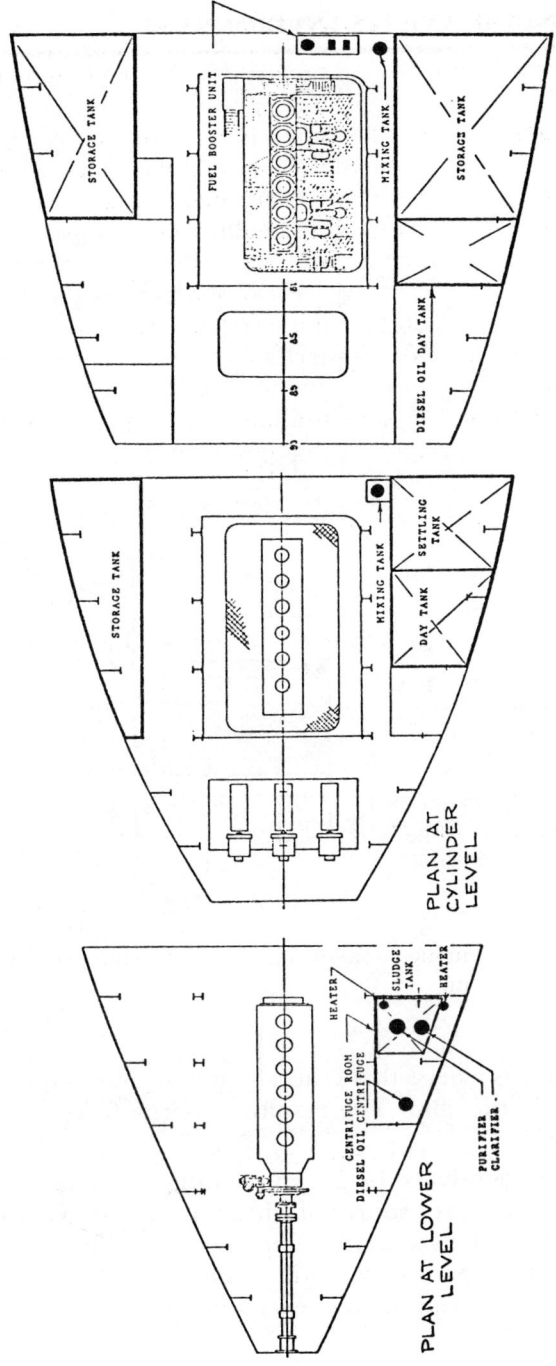

Figure 12.1 Typical engine room arrangement showing principal components of the fuel system.

238 THE ENGINE AND ITS ENVIRONMENT

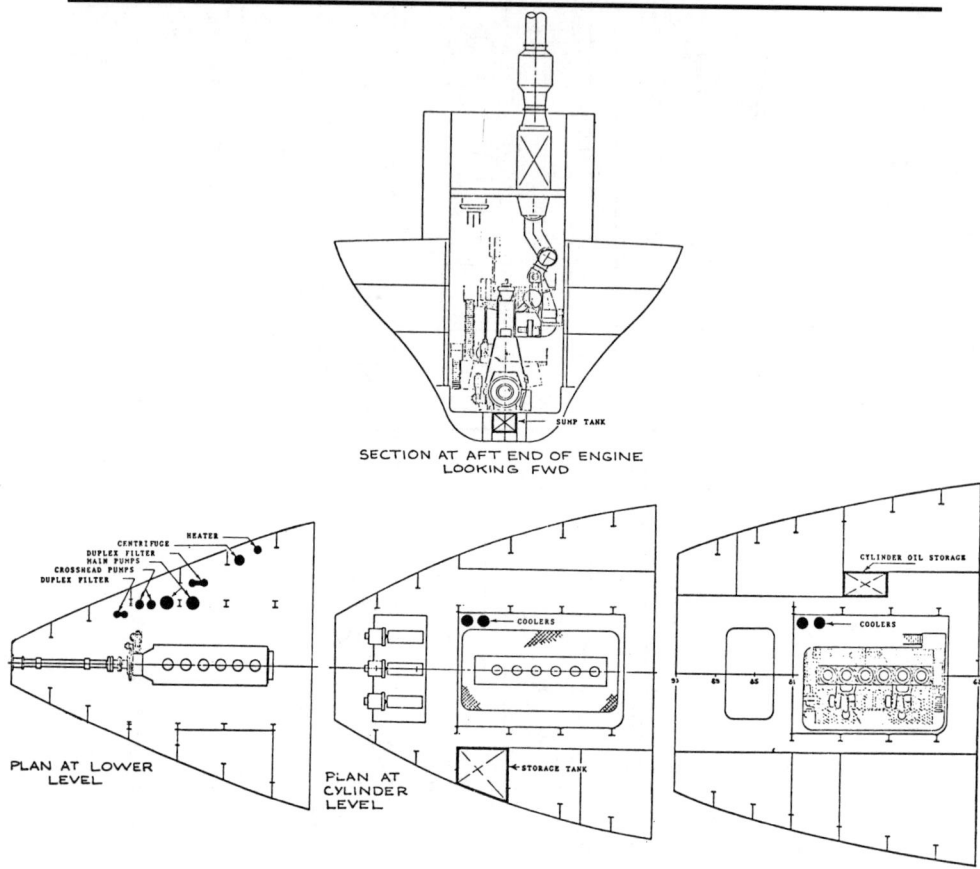

Figure 12.2 Typical engine room arrangement showing principal components of the lubrication system.

3. The engine is a mass that is accelerated by motions (rolling, pitching, etc) of the ship. The resulting inertia forces bear upon the foundation.
4. The torque produced by a reciprocating engine is not steady, but cyclic, and hence is a source of vibration excitation to the surrounding structure. This structure must be of rigidity sufficient to keep its lowest frequency of vibration well above the frequency of any major engine (or propeller) excitation.

Typical structural arrangements and scantlings that result from these considerations are pictured in Figure 12.7 (Sulzer dwg 9-107.095.035).

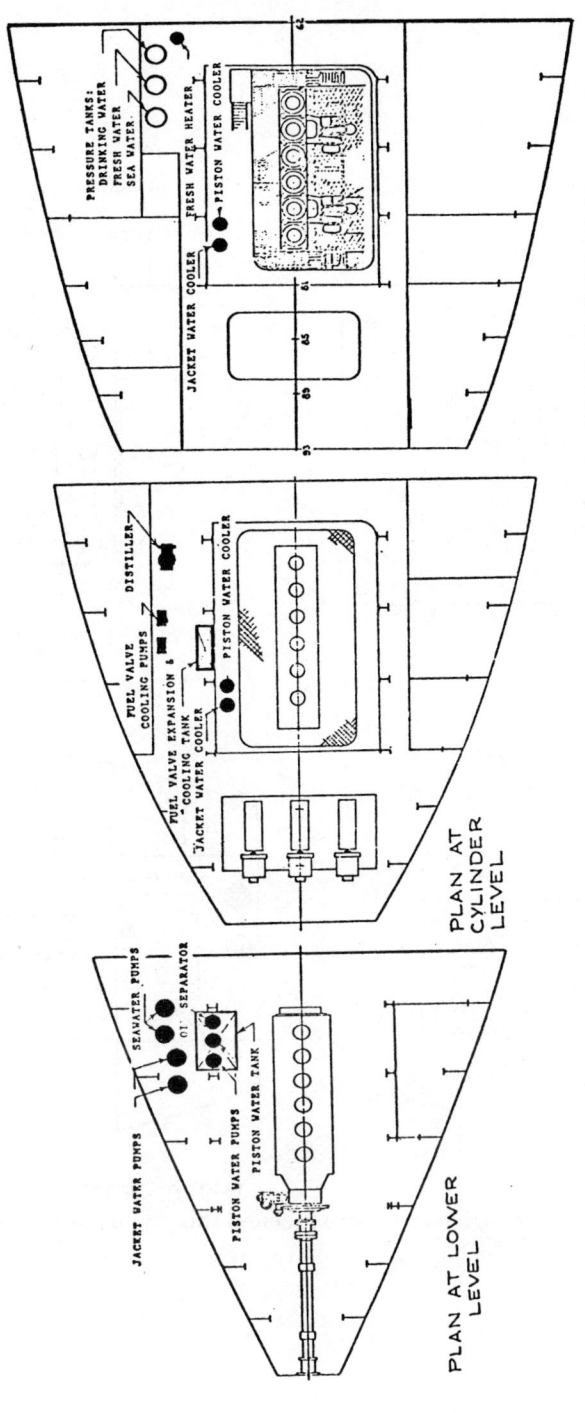

Figure 12.3 Typical engine room arrangement showing principal components of the engine cooling water systems (some domestic water components appear in the rightmost sketch).

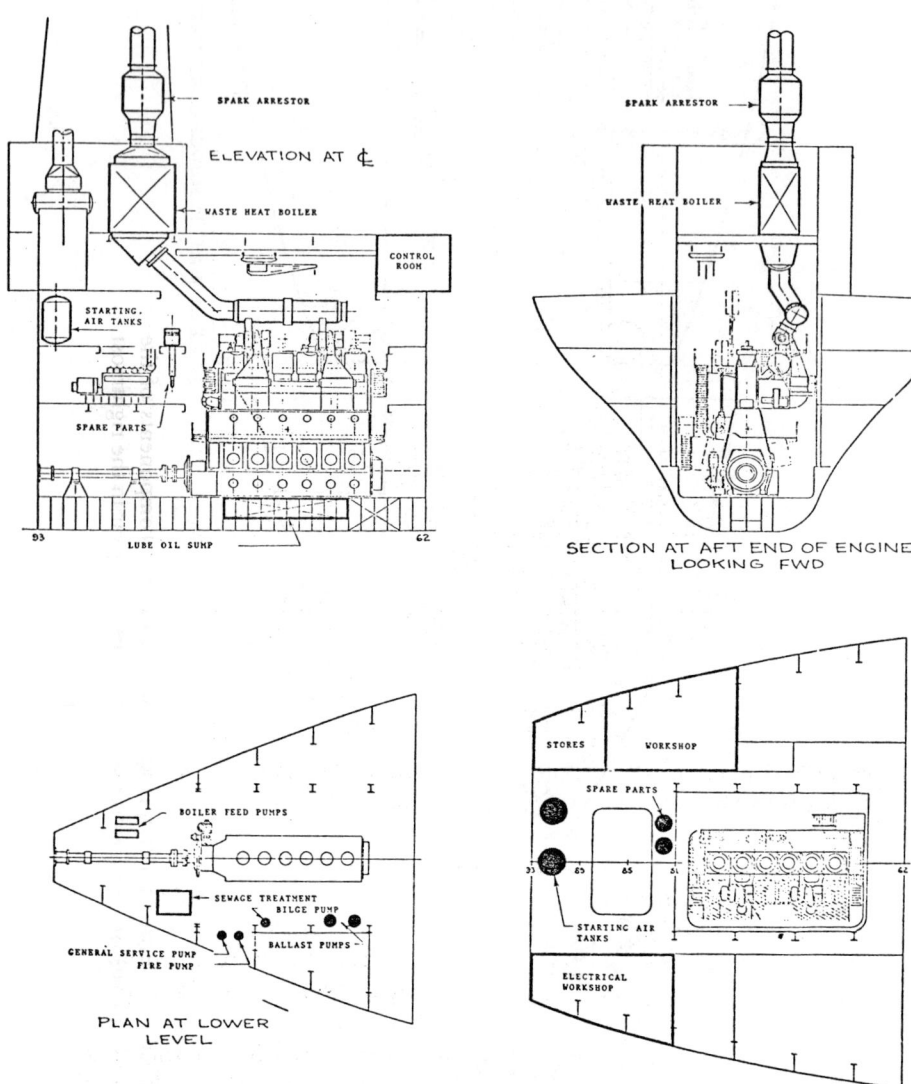

Figure 12.4 Typical engine room arrangement showing miscellaneous auxiliary components.

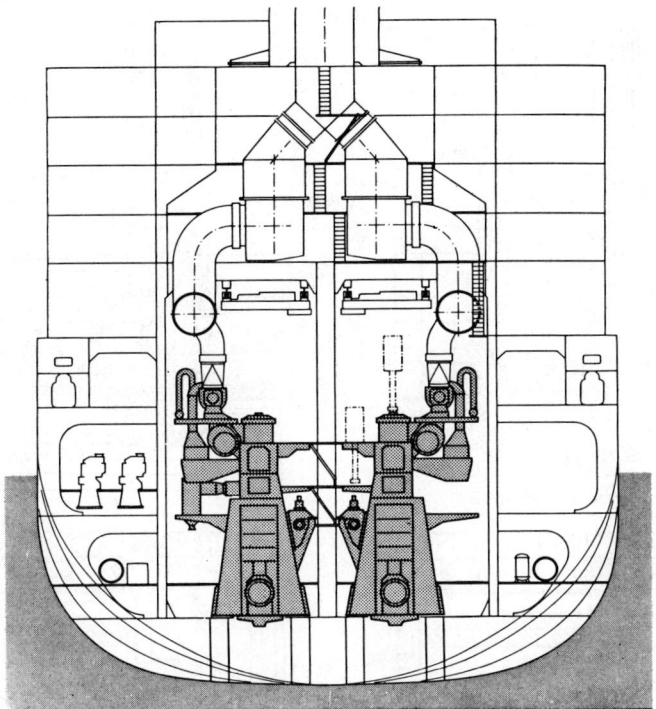

Figure 12.5 Engine room section with twin Kawasaki-MAN engines of 40,000 bhp each (from *Shipbuilding & Marine Engineering International*).

Cross sections of bottom structure as recommended by Sulzer for its several engine models are shown, with plate thicknesses in millimeters. Recall from Chapter 10 and from Figure 12.2 that the lube oil sump tank lies below the Sulzer engines. This tank is evident in Figure 12.7, though not labeled.

Figure 12.8 shows details of the inner bottom plating at the area of contact with the support flange of the engine bedplate (three alternative schemes recommended by Sulzer are given). The essential element is a machined surface with the slight slope shown, the slope being an aid to fitting support chocks between the machined surface and the bedplate flange.

The chocks are small squares of steel or cast-iron plate of a precise thickness. In the usual erection sequence, the bedplate, with crankshaft in place, is carefully leveled in a shore workshop. It is then lowered into the ship, and supported temporarily by jackscrews (illustrated by Figure 12.9) bearing on the machined inner bottom area. The jackscrews are

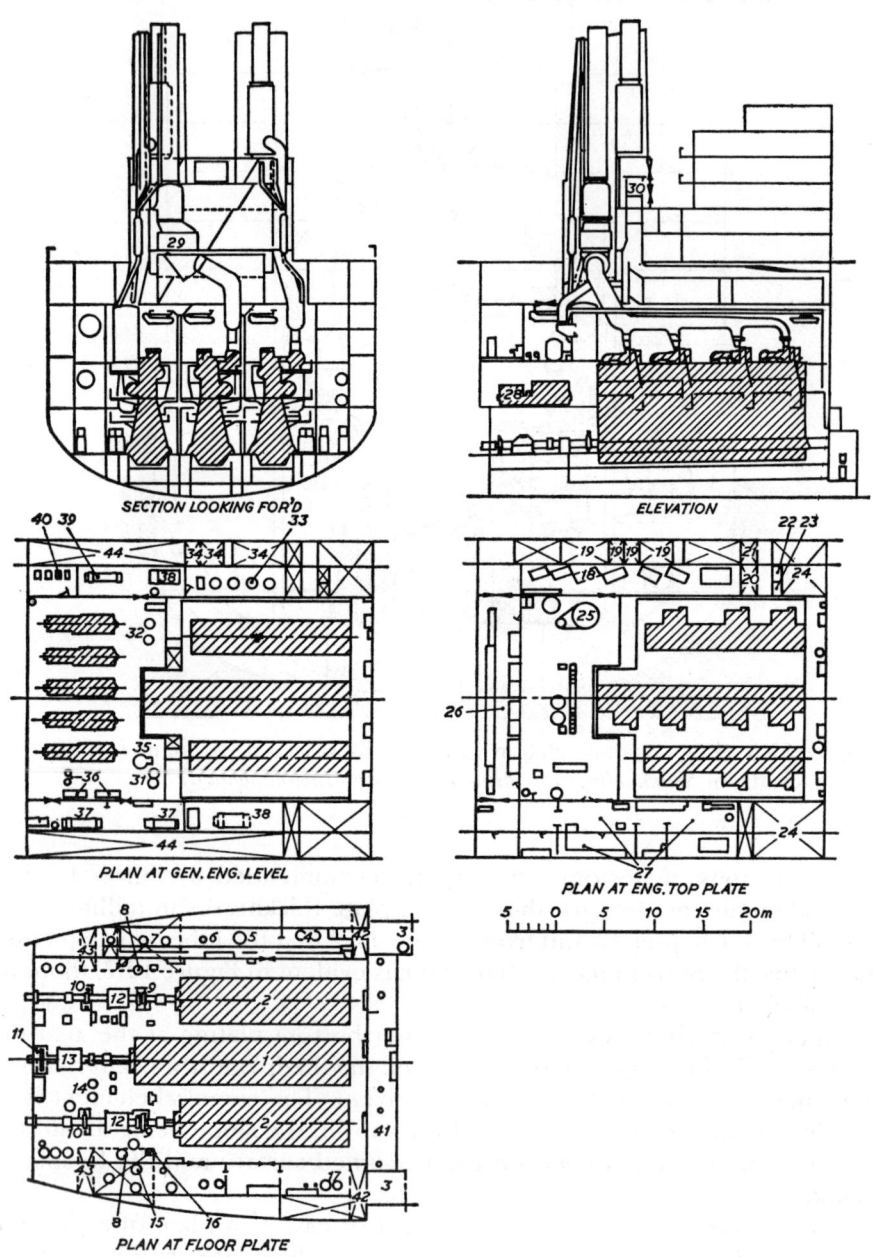

adjusted until the leveling measurements made ashore are reproduced. After leveling, the erection of the remainder of the engine upon the bedplate proceeds. At some point, usually after engine erection is complete and the ship is afloat, any adjustment necessary (as for restoring bedplate level, or for aligning engine output coupling to propeller shaft) in the jackscrews is made. The gaps between bedplate and inner bottom plate are then measured, and the chocks are machined to suit and fitted to support the engine in exactly the position established with the jackscrews.

The use of chocks cast in place, using epoxy resin as the casting material, should be mentioned here. This technique has the obvious appeal of avoiding the careful measuring, machining, and fitting that is necessary with the connventional chocks, and they seem to be widely used, especially in repair work. However, some engine builders express skepticism over their long-term durability, and refuse to endorse them for their engines.

Permanent hold-down bolts pass through bedplate flange, chock, and

Figure 12.6 Machinery arrangement with triple low speed engines [from Kongsted (1975)].

1 Center engine (12K84 EF)
2 Wing engines (9K84 EF)
3 Stabilizer compartments
4 Bilge pumps
5 Bilge water separator
6 Sludge pump
7 S.W. pumps W/E
8 L.O. pumps W/E
9 Hydr. clutch W/E
10 Disk brake W/E
11 Disk brake C/E
12 Thrust blocks W/E
13 Thrust blocks C/E
14 L.O. pumps C/E
15 S.W. pumps C/E
16 L.O. filters
17 Fire and deck wash pumps
18 Starting air comp.
19 L.O. storage tanks
20 Diesel oil serv. tank C/E
21 Diesel oil serv. tank W/E
22 Heavy fuel serv. tank W/E
23 Heavy fuel serv. tank C/E
24 Heavy fuel settling tanks
25 Oil-fired boiler
26 Engine control room
27 Workshops and stores
28 Diesel alternators (826MTBH-40)
29 Exh. gas economizer
30 Eng. room ventilators
31 F.W. pumps C/E
32 F.W. pumps W/E
33 Fuel oil purifiers
34 L.O. storage tanks
35 F.W. evaporator
36 F.W. pumps and coolers for generator eng.
37 F.W. and L.O. coolers C/E
38 L.O. coolers W/E
39 F.W. coolers W/E
40 Transformers
41 Pump room for fuel transfer and ballast pumps
42 Forward sea chests
43 Aft sea chests
44 Water ballast

ENGINE SEATINGS

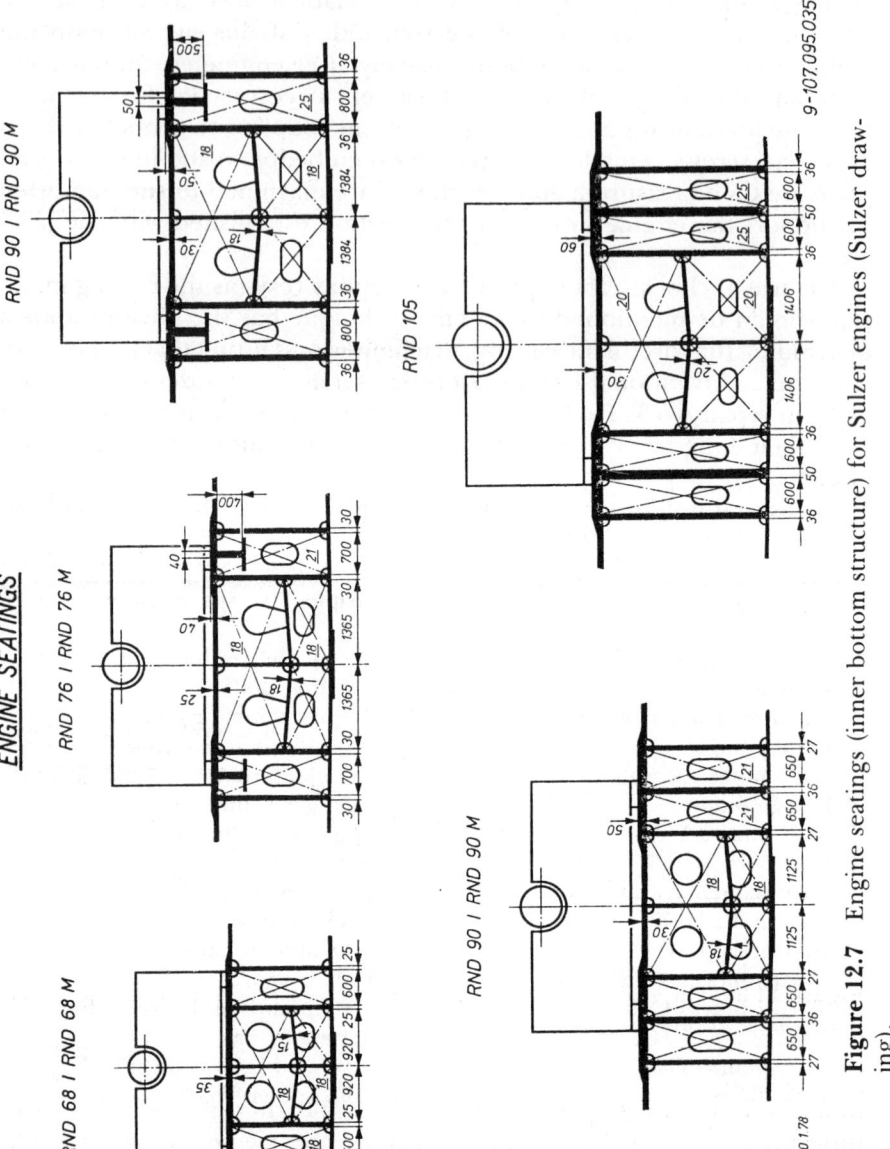

Figure 12.7 Engine seatings (inner bottom structure) for Sulzer engines (Sulzer drawing).

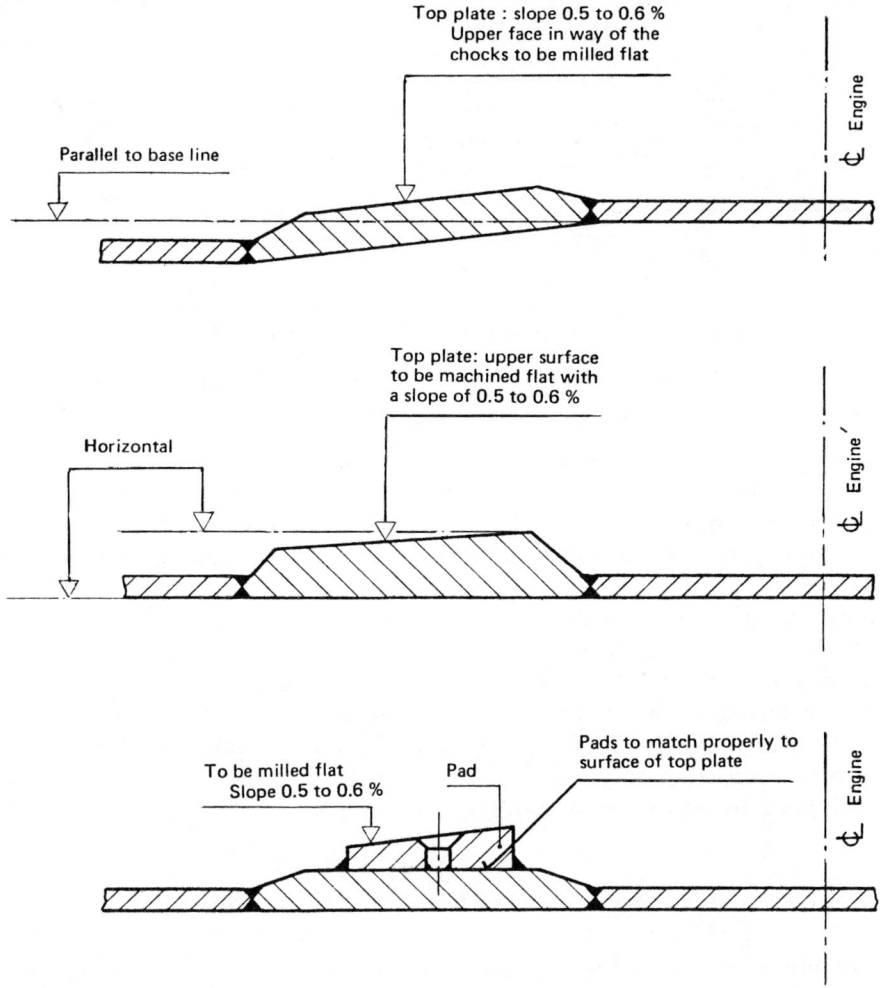

Figure 12.8 Three alternatives for inner bottom plating in way of engine bedplate (Sulzer drawing).

inner bottom.* In the vicinity of the thrust bearing, the bolts are fitted (holes reamed after drilling, bolts machined to precise fit), this for adequate transmittal of the thrust between engine and inner bottom. Elsewhere, ordinary bolts (holes drilled only, bolts to standard size) are used.

---

*Some means of access to the lower ends of the bolts must be provided, although this detail is not evident in any of our figures.

246 THE ENGINE AND ITS ENVIRONMENT

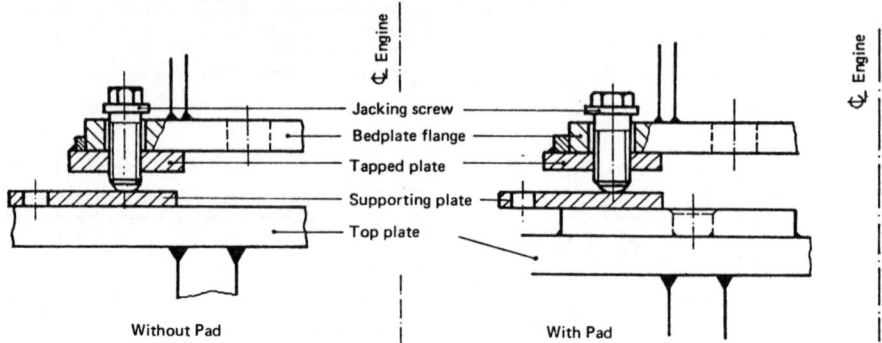

Figure 12.9 Jackscrews to support engine bedplate during erection (Sulzer drawing).

The minute sliding thereby allowed is to accommodate differential thermal expansion between bedplate and inner bottom.

The engine is usually secured against athwartship sliding by brackets welded to the inner bottom, and bearing on the side of the bedplate bottom flange via chocks.

Additional engine support is sometimes provided by sway braces connecting the engine at the level of the top of the cylinder (typically) to nearby structure. Their purpose is to provide rigidity against lateral vibration of the engine. (See Section 12.7.) Usually they are omitted in original construction, and installed only if this vibration is later experienced to a troublesome degree.

12.4 TORSIONAL VIBRATION OF ENGINE AND SHAFT

Torsional vibration of the rotating parts of an engine and its connected power transmission components is one of the more serious hazards threatening the success of a propulsion plant design. It can occur with any type of engine, since uneven wake distribution around the propeller can cause that ubiquitous component to be a source of torque excitation, but its occurrence with a reciprocating engine is much more likely because of the strongly fluctuating torque produced by each engine cylinder. A universal practice in marine engineering design, in consequence, is to perform a torsional vibration analysis of the engine-shaft system.

Now, the full story of torsional vibration and its analysis is quite too much for this book. I shall have to assume that your background includes a course in vibration theory, and here offer only a summary of the situation intended as a review, and to mention points unique to the engine we are looking at. The torque excitation phenomena that are unique to

TORSIONAL VIBRATION OF ENGINE AND SHAFT

reciprocating engines are often not covered in vibration courses, nor are the calculations of engine spring constants and inertias, so you may not have encountered them before, but they can be better studied in specialized sources [den Hartog (1956)]; hence this topic is only summarized here also.

An analysis of engine-shaft torsional vibration typically consists of calculation of natural frequencies and mode shapes of the vibrating assembly, calculation of the frequencies and magnitudes of the torque excitation, and calculation of the deflections and stresses that result.

The first of these calculations is based on modeling the rotating system as a series of circular disks mounted on a straight elastic shaft. Key parameters are mass moments of inertia of the disks, and torsional spring constants of the shaft sections between disks. The usual model includes an equivalent disk for each assembly of piston, piston rod, crosshead, connecting rod, and crank throw. Because the collective geometry of these items is complex, and because their motion is a combination of rotation and reciprocation, an accurate reduction to the equivalent disk is no simple process. Calculation of the torsional spring constant of a section of crankshaft is also difficult. If simple formulas would suffice we would have them here, but they don't, so you must seek the information in those specialized sources [den Hartog (1956)].

The result of the first calculation is a set of natural frequencies of vibration (one for each degree of freedom* of the system), and a list of amplitude ratios for each of these frequencies. If an exciting torque coincides with one of the frequencies, the corresponding vibration occurs at that frequency, and the vibrational amplitude of each equivalent disk is the value given by the corresponding amplitude ratio. The system is then said to be vibrating in one of its *principal modes*. For example, if vibrating at its lowest natural frquency, it would be said to be in "mode one" or in its "first mode"; if the next highest frequency predominates, the system is said to be in its "second mode"; and so on.

Table 12.1 offers as an example a summary of key data and results from just such a torsional analysis. The ship is a bulk carrier powered by a low speed engine of six cylinders, 760 mm bore, 12,000 hp at 122 rpm. The engine is located aft, with a shaft diameter of 490 mm, propeller shaft diameter of 590 mm. The table gives the mass polar moments of inertia of the equivalent disks, the spring constants of the shaft sections between disks, and the resulting frequencies and mode shapes for the first two modes.

Figure 12.10 (MAN Fig. 36 36 421 W) represents the first two vibration

---

*"Degrees of freedom" here is one less than the number of disks used in modelling the rotating system.

TABLE 12.1 EXAMPLE SUMMARY OF A TORSIONAL VIBRATION ANALYSIS

Engine Data		Shaft Data	
Number of cylinders	6	Crank journal	600 mm
Bore	760 mm	Crank pin	600 mm
Stroke	1550 mm	Intermediate shaft	490 mm
Brake horsepower	12000	Propeller shaft	590 mm
Rpm	122		
Mean effective pressure	1150 kPa		

Mass Number	Moment of Inertia (N cm s^2)	Spring Constant (N cm/rdn)	Amplitude Ratio[a]
1	0.0141×10^6	0.1005×10^{12}	1.0000
2	0.5171×10^6	0.0766×10^{12}	0.9997
3	0.5171×10^6	0.0766×10^{12}	0.9839
4	0.5171×10^6	0.0766×10^{12}	0.9529
5	0.5171×10^6	0.0766×10^{12}	0.9073
6	0.5171×10^6	0.0766×10^{12}	0.8478
7	0.5312×10^6	0.0445×10^{12}	0.7752
8	0.7523×10^6	0.0144×10^{12}	0.6291
9	0.0341×10^6	0.0092×10^{12}	0.1041
10	0.0429×10^6	0.0192×10^{12}	−0.7198
11	2.97×10^6		−1.1117

Mass 1   Forward part of shaft
Masses 2–7 Cylinder components
Mass 8   Flywheel + shaft
Mass 9   Shaft coupling
Mass 10   Shaft coupling
Mass 11   Propeller + entrained water

Natural frequency 7.6 Hz (first mode)

[a] Ratios for first mode.

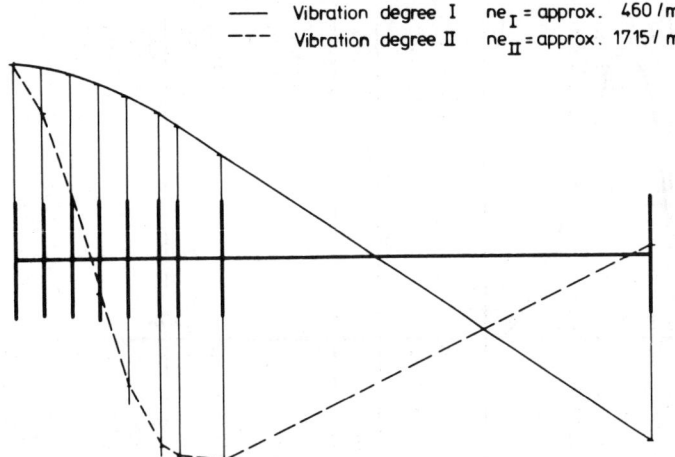

Figure 12.10 Typical mode shapes for first two torsional modes (MAN drawing).

modes of a MAN K6SZ90/160A engine connected to a typical shaft and propeller. (Not the same engine or ship as Table 12.1, however, but quite similar.) The vertical bars along the shaft line are, reading from left to right, the equivalent disks representing the moving parts of the six cylinders, the camshaft drive, the barring wheel,* and the propeller. The two curves are plots of the amplitude ratios, these being the ratios of vibrational amplitude at each disk to the amplitude at the forward cylinder. The curves therefore constitute the "mode shape" for each of the two modes plotted.

No actual magnitudes are shown in Figure 12.10, so that the stresses resulting from the relative vibrational displacements are not evident, but the shapes of these curves are typical of the first two modes of engine-shaft-propeller systems, and significant general information is revealed by them. The slope of the mode-shape curve is a measure of the magnitude of vibrational stress (we'll assume that the figure here is drawn to scale). For the first mode, this region of greatest hazard occurs in the shaft; hence any failure resulting from excessive stress is likely to afflict that component. The greatest second mode slope occurs in the crankshaft, putting it at hazard from that mode. Note also the large displacement (relatively) between ends of the crankshaft. Camshaft drive-gear

---

*The toothed wheel by which the "turning gear" or "barring gear" motor rotates the engine when this is needed during maintenance, etc.

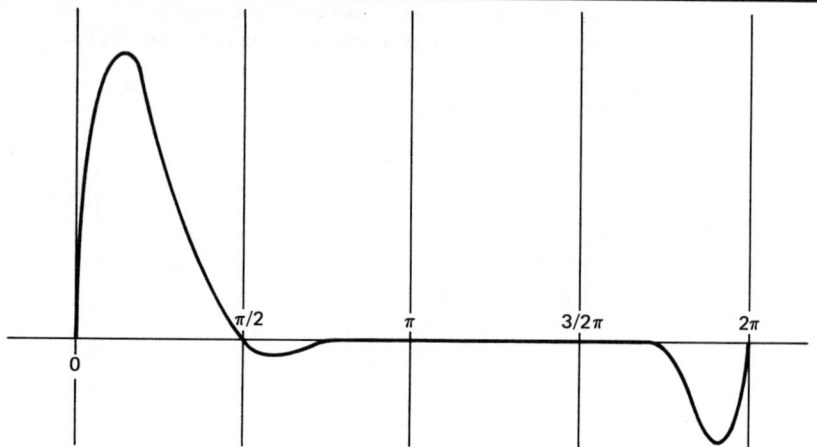

Figure 12.11 Typical torque curve (per cylinder, one rotation).

teeth, for example, may be given excessively large dynamic loads by this phenomenon.

Now, the principal exciter of torsional vibration is the fluctuating torque of the engine. At each cylinder the torque passes through a complete cycle (including a negative portion during compression) for each rotation of the crankshaft. There is thus a cyclic torque for each cylinder at the frequency of engine rotation. See Figure 12.11 for an example. In addition, the many harmonics of this cyclic torque curve are present.* Each such harmonic constitutes an "order" of excitation, this term denoting the ratio of frequency to engine rotational frequency. For example, "third-order excitation" refers to excitation by the torque harmonic that has a frequency three times engine rotational frequency.

There are many orders present, and likewise many natural frequencies of the torsional vibration. When any order coincides with any frequency, a resonance occurs that may cause dangerous vibrational (and hence stress) magnitudes. But that conditional word "may" is important, for a mere coincidence of frequencies is not sufficient to produce a resonance of great magnitude. The reasons can be seen by examination of the equation (borrowed from [den Hartog (1956)]) that expresses the rate of energy transfer from engine source to vibrating component:

$$\text{energy transfer} = \pi\, M_n \beta_n \sin \phi_n \qquad (12.1)$$

*I'm assuming that you are familiar with the concept of harmonic analysis by which any cyclic function is broken up into its sinusoidal components.

where M_n = magnitude of the torque harmonic
  B_n = magnitude of the vibration (amplitude ratio)
  ϕ_n = phase angle
  n = index for the particular cylinder

The value of M_n is not uniform for all of the many torque harmonics, but decreases strongly for the higher ones. Figure 12.12 (MAN Fig. 36 36 329/3V) illustrates this point with curves of harmonic magnitude (labeled "tangential excitation factors") for MAN engines operating at five values of mean effective pressure.

The value of β_n is proportional to the amplitude ratio, hence may have a distinctly different value for each cylinder (recall Figure 12.10).

Now, consider the phase angle ϕ_n. If this could be zero, the engine obviously could not excite the vibration. Wonderful, indeed, but there is no way of obtaining this benificent state, nor even of knowing what the phase angle is—the only safe assumption in design is that it will be the worst value possible, 90 degrees. However, the value cannot be the same at all cylinders because they do not fire simultaneously. For example, when concerned with the second-order excitation in a six-cylinder engine (six firings per revolution, two vibration cycles per revolution, typical firing order 1-6-2-4-3-5), you will find three equally spaced firings per

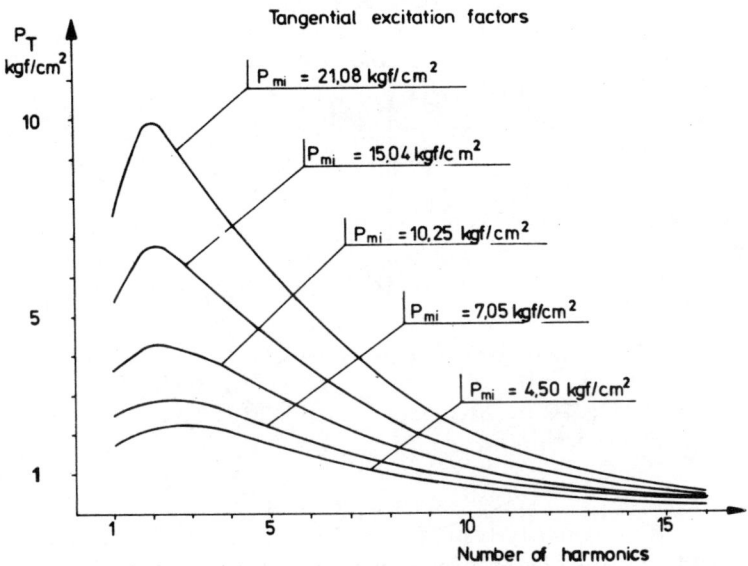

Figure 12.12 Typical magnitudes of harmonics of engine torque as function of mean effective pressure (MAN drawing).

vibration cycle, and by plotting the product $M_n B_n$ as a phasor diagram, you produce Figure 12.13 (the amplitude ratios of Figure 12.10 are used for β—no actual units, though). It is the *resultant* of the six vectors shown that must always be conservatively assumed to have the 90 degree phase angle, but note that the resultant magnitude depends not only on the magnitudes of M_n and B_n, but on the angles within the diagram, and hence on the order number, on the number of cylinders, and on the sequence in which they fire. In general, then, this diagram must be drawn for each order and each mode to determine which combinations produce major excitations.

The accelerations of the reciprocating masses at each cylinder produce an inertia torque (see Section 12.11). Although small compared to the gas-pressure torque, it is not negligible. M_n in equation (12.1) therefore is assumed to include contributions from both gas pressure and inertia.

The third aspect of the torsional vibration analysis is calculation of the actual magnitudes of vibrational deflection and the resulting stresses. The essence of this is finding the level of deflection at which energy input from the excitation equals the energy dissipated by damping. The damping energy is usually considered to come from three sources, namely:

1. Propeller damping caused by its vibrational displacement relative to the water.

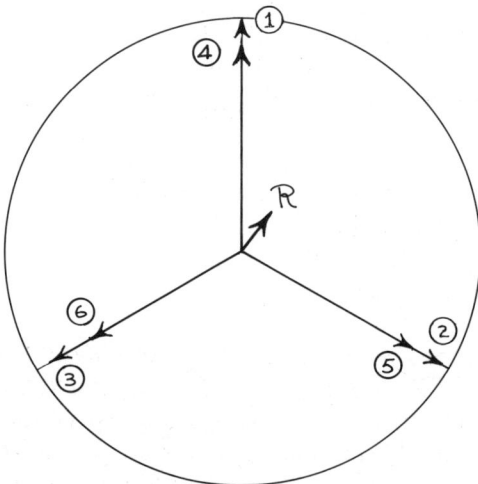

Figure 12.13 Typical phasor diagram for engine torsional excitation [each vector is $M_n B_n$ in equation (12.1)].

2. Damping by elastic hysteresis and sliding contacts with viscous lubricant.
3. Damping by the engine (i.e. piston working against gas pressure).

The first two of these are discussed in Chapter XI of MARINE ENGINEERING [Long (1971)], where you will find propeller damping energy given by

$$E_p = \pi \left[2 - \frac{J}{K_Q} \frac{\partial K_Q}{\partial J} \right] \frac{Q}{\Omega} \omega \Theta^2_p \qquad (12.2)$$

and the hysteresis/sliding damping energy by

$$E_I = \Sigma \frac{1}{2} \alpha \omega^2 I_n \Theta^2_n \qquad (12.3)$$

(summation over n cylinders)

Engine damping is discussed in several sources [Wilson (1956), Salzman and Pamidi (1973)]. The latter gives engine damping energy as

$$E_E = \pi K_e \omega \Sigma I_n^{0.8} \Theta_n^{0.8} \qquad (12.4)$$

(summation over n cylinders)

In the above three equations, symbols are defined as follows:
J = propeller advance coefficient
I_n = mass polar moment of inertia of the nth equivalent disk
K_e = coefficient in the range 12 to 40
Q = propeller torque
K_Q = propeller torque coefficient
Θ_n = torsional displacement at nth equivalent disk
Θ_p = torsional displacement of propeller
Ω = propeller speed of rotation
ω = vibrational frequency

Figure 12.14 (MAN Fig. 36 36 423W) is a plot of the stress as a function of rpm in the shaft of a propulsion system using the MAN K6SZ 90/160A engine, and resulting from interaction of the first vibrational mode with the sixth-order engine harmonic. Curves of stress allowed by one of the classification societies—Lloyds Register of Shipping—are shown on this figure; the lower of the two is the limit for continuous operation, and the other is for momentary ("passing through") operation. The stress plot

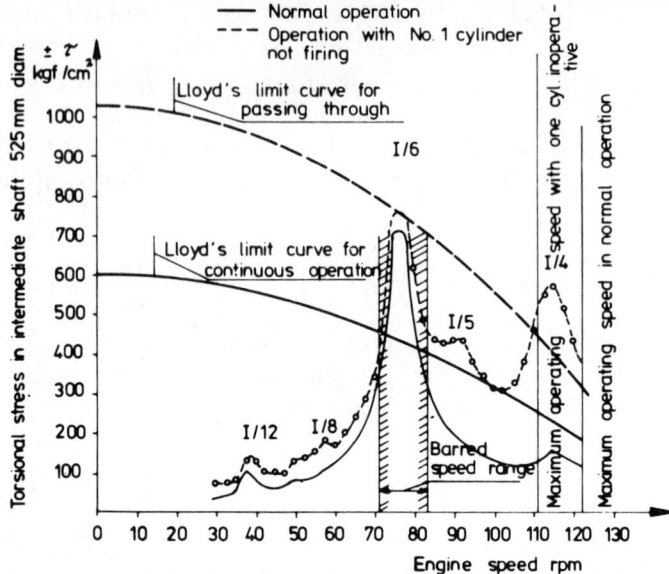

Figure 12.14 A typical torsional resonance: torsional stress in propulsion shafting as a function of rpm (MAN drawing).

exceeds the lower of these curves, but not the upper, and thereby suggests an ameliorative technique that is quite common: the barred speed range. In this instance the shaft stress exceeds the continous limit between 71 and 83 rpm; hence this range of rpm is "barred" or forbidden for anything but passing to and from the normal full-power operating range of speeds.

The barred speed range is often used in the situation pictured by Figure 12.14 since its cost is only inconvenience. If it is unacceptable, as when the momentary stress limit is exceeded also, or when continuous use of the barred rpm will be necessary for the ship, measures such as addition of a torsional damper* to the engine, increase in shaft diameter to raise the vibrational frequency, etc, must be used. An excellent paper by Spaetgens [Spaetgens (1961)] describes cases in which cures of several types were efficacious.

*If you are not familiar with dampers, see Figure 12.15, which illustrates one of the several types of dampers used with diesel engines. It consists of an central member rigidly connected to the engine crankshaft, and a concentric outer member joined to the inner by springs. The outer member therefore can vibrate with a large amplitude, and hence a lot of energy is dissipated viscously by the oil that fills the damper.

AXIAL SHAFT VIBRATION

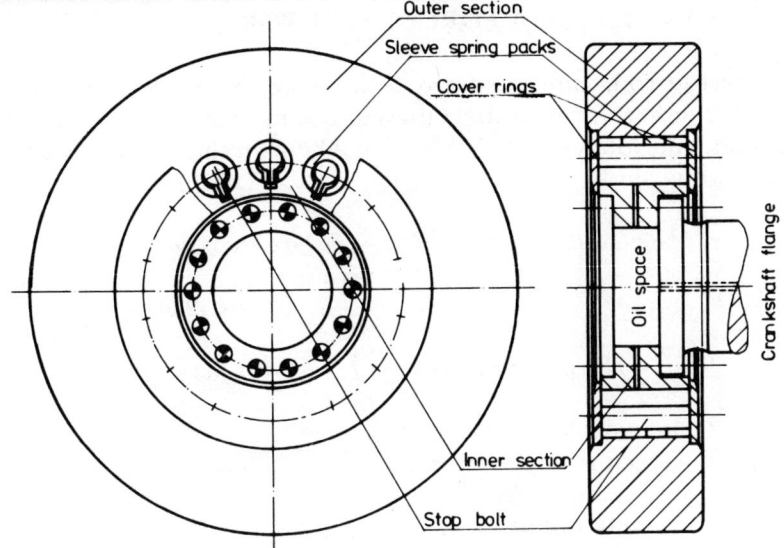

Figure 12.15 A vibration damper (MAN drawing).

Note that the stress curve of Figure 12.14 is repeated for the situation in which one cylinder is not firing (as the result of a failure, no doubt). Since the absence of the gas-pressure torque distorts the resultant in a phasor diagram such as Figure 12.14, the curve must be somewhat different. Because of the possibility of casualty operation of an engine, this additional analysis is commonly performed.

12.5 AXIAL SHAFT VIBRATION

Axial vibration of the shaft is usually of minor concern to the designer of a low speed engine propulsion plant, since it is not the ubiquitous threat that torsional vibration is. However, it can occur through excitation by coupling between torsional and axial motions of the crankshaft. Of course, the crankshaft has no deliberate axial motion, but connecting rod thrust on a crankpin produces bending in the webs of the crank throw, and this in turn produces an axial displacement of the shaft as a whole. If a resonance does indeed occur with a troublesome magnitude, then the likely cure is an external damper added to the forward end of the crankshaft.

12.6 ENGINE INERTIA FORCES AND TORQUES

Piston, piston rod, and crosshead reciprocate, hence accelerate continously in the vertical direction; the connecting rod reciprocates at its upper end, and rotates at its lower end. The resulting inertia forces per cylinder are given by den Hartog [den Hartog (1956)] as

$$\text{Vertical forces} = (\dot{m}_{rec} + \dot{m}_{rot})r\omega^2 \cos(\omega t) \quad (12.5)$$

$$+ \dot{m}_{rec}\frac{r^2}{l}\omega^2 \cos(2\omega t)$$

$$\text{Horizontal forces} = \dot{m}_{rot}r\omega^2 \sin(\omega t) \quad (12.6)$$

where l = stroke length
$\dot{m}_{rec}$ = reciprocating mass
$\dot{m}_{rot}$ = rotating mass
r = crank radius
t = time
ω = rotational frequency (2 × rpm/60)

The forces at frequency t are called "primary" forces, while those at 2t are called "secondary."

The technique of counterbalancing the crankshaft, which places a mass equivalent to the connecting rod rotating mass on the opposite side of the center of rotation, can make m$_{rot}$ zero, and so eliminate the horizontal forces, but because it is not feasible to counterbalance reciprocating mass, the vertical inertia force per cylinder is always present. Nonetheless, forces—and the moments that are the consequence of forces—can be balanced by cancellations among cylinders. To illustrate this point, consider a four-cylinder engine (cranks at 90 degrees, firing order 1-2-4-3), and note Figure 12.16, which is a set of phasor diagrams showing the phase relationships among the vertical inertia forces and resulting moments for this engine.

In Figure 12.16, the first two phasor diagrams show that the resultant primary and resultant secondary forces are zero. On the other hand, each force produces a moment about any arbitrarily chosen point. To illustrate the consequences, let's choose cylinder 1 as that point, and multiply the length of each vector in the first two diagrams by its distance from the centerline of that cylinder. The result is the second pair of phasor diagrams, and you note that the moments resulting from the primary vertical forces have a nonzero resultant. The four-cylinder engine thus has an unbalanced primary moment tending to bend it vertically at the primary frequency, that is, the frequency of shaft rotation.

ENGINE INERTIA FORCES AND TORQUES   257

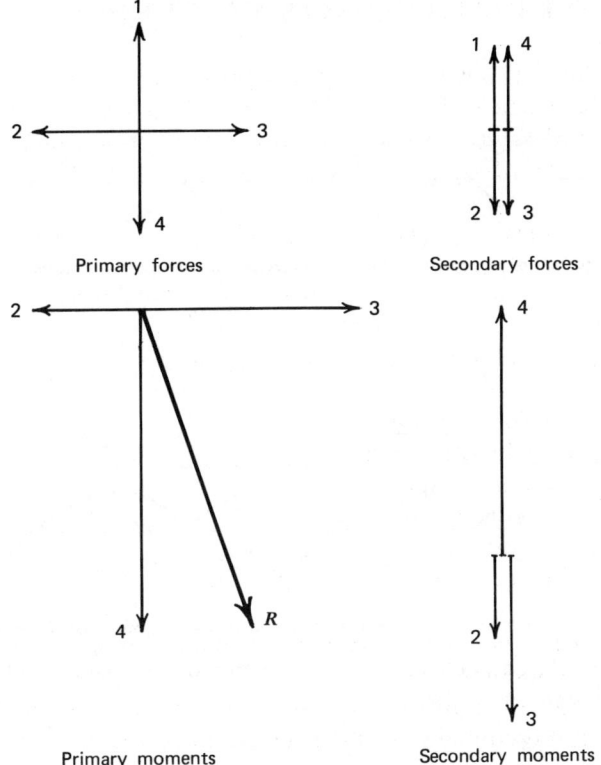

Figure 12.16 Phasor diagrams to show balance or unbalance of engine forces and moments.

The phasor diagrams of Figure 12.16 obviously depend on the number of cylinders, so that the cancellation or noncancellation of inertia forces and moments depends on this number, and with respect to horizontal forces and moments, must also depend on whether m_{rot} is counterbalanced to zero.

These forces and moments must obviously be resisted by the structure of the engine, and if that structure were infinitely stiff, no consequences would be evident external to the engine. Engine structure isn't infinitely stiff, of course, so there is the possibility of a cyclically bending engine exciting the hull. Whether it does depends on the resonance of hull frequency and excitation frequency, and on the location of the exciter (the engine) with respect to nodes and anti-nodes of the hull vibration. See Figure 12.17 to illustrate the latter point, and also see Section 12.7 for further points.

258   THE ENGINE AND ITS ENVIRONMENT

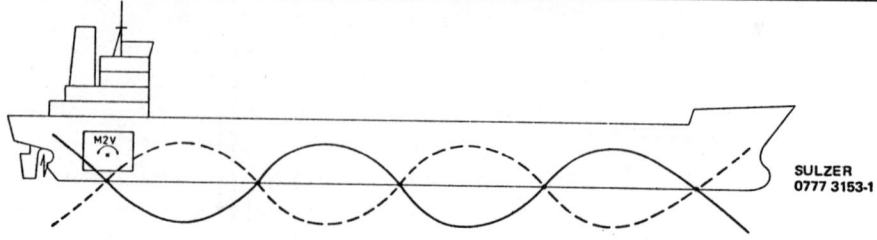

Example of engine position which can promote hull vibration (schematic).

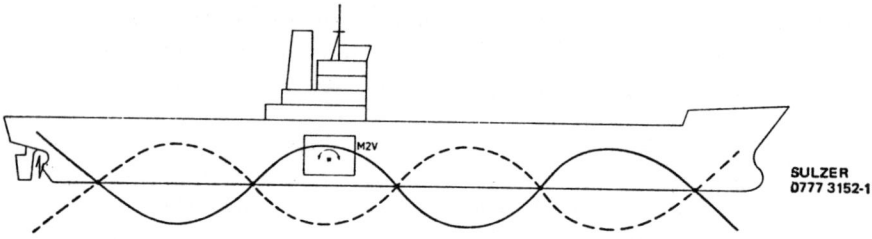

Example of engine position which cannot promote hull vibration (schematic).

Figure 12.17   To show effect of engine position on its excitation of hull vibration (Sulzer drawing).

Section 12.3 has noted that inertia torque adds to the gas pressure torque as an exciter of shaft torsional vibration. The forces expressed by equations (12.2) and (12.3) produce this torque; the torque equation from den Hartog is

$$M = \frac{1}{2}\, \dot{m}_{rec}\, \omega^2 r^2 \left(\frac{r}{2l} \sin(\omega t) - \sin(2\omega t) - \frac{3r}{2l} \sin(3\omega t) \right) \quad (12.7)$$

This is the usual approximation used, an approximation because it omits all terms beyond order three as being of insignificant magnitude. Symbols are the same as those used in equations (12.5) and (12.6).

12.7   ENGINE LATERAL EXCITATIONS

The vertical force produced by gas-pressure on the piston crown is, of course, transmitted to the crankshaft by the piston rod, crosshead, and connecting rod. Because this last component transmits a force at a cyclically varying angle to the vertical, it produces a horizontal component of force on crosshead bearings and crankshaft bearings. This force acts in opposite directions on the respective bearings, and so cancels internally, but produces cylinder-to-cylinder moments in the manner of the inertia forces discussed in the previous section. The result may be a transverse bending vibration of the engine, with many orders present, just as with the torsional excitation. Hull vibration may in turn be excited, just as in the instance of the excitation discussed previously.

The use of lateral sway bracing for engine support was mentioned in Section 12.2. By providing increased stiffness, this support raises frequency of response away from the major orders of excitation. It is the usual cure for vibration excited by the excitation just discussed.

12.8 SUMMARY OF DYNAMIC PROBLEMS

The discussions in Sections 12.3 through 12.6 have noted that vibrations and the excitations thereof depend on characteristics of propeller and shaft, on number of cylinders, on location of the engine, and on the reciprocating and rotating masses within the engine. Some of these depend on the design of the ship; hence every unique ship design can have unique responses. Nonetheless, generalizations are possible, giving one a preliminary idea of what to expect from a particular design.

An example of such a generalization is given by Table 12.2. This table is furnished by Sulzer for its RND..M engines, and neatly summarizes typical results from the entire spectrum of those engines. Note, however, the caveat below the table: for guidance only.

12.9 NOISE

There's no doubt that propulsion machinery is noisy, and that noise, generally, is a bad thing. Among prominent and well-recognized effects are its interference with conversation and its irreversible damage to human hearing (for long exposure at high levels).

Beginning in the neighborhood of 1960, noise aboard ship began to attract the interest of regulatory bodies, and recommendations, guidelines, and a few mandatory regulations have appeared since then. And it seems likely that recommendations and guidelines will progress into regulations as marine designers gain experience in working to suit them. An example is given by Table 12.3, the sound levels proposed (such was

TABLE 12.2 SULZER'S SUMMARY OF DYNAMIC CHARACTERICS

Number of Cyl.	Axial Vibration	Engine Vibration Mode — Second Order Vibration from Unbalanced Couple	Transverse Vibration	Shaft Line Torsional Vibration — Aft-Installation (Very Short Shaft Line) RND-M			Shaft Line Torsional Vibration — Semi-Aft Installation (Medium Length Shaft Line) RND-M			Shaft Line Torsional Vibration — Mid-Ship Installation (Very Long Shaft Line) RND-M			Number of Cyl.
				68	76	90	68	76	90	68	76	90	
4	No special requirements	Engine room space for subsequent mounting of a second order balancing gear should be provided		No barred speed range 1)	—		Barred speed range (critical 4th Order)	—		Barred speed range (critical 4th order)	—		4
5				No barred speed range 1)	—		Barred speed range (critical 5th order)	—		Barred speed range (critical 5th order)	—		5
6			Provision for subsequent installation of transverse stays recommended	Barred speed range (critical 6th order)	Barred speed range (critical 7th order)		Barred speed range (critical 6th order)			Barred speed range (critical 6th order 2)	No barred speed range		6
7	Countermeasure may be found necessary. Please consult engine builder						No barred 2) Barred speed range	Barred speed range (critical 7th order)		No barred speed range			7
8	No special requirements			Barred speed range (critical 8th order)			No barred speed range 2)			No barred speed range			8
9	Countermeasure may be found necessary. Please consult engine builder			No barred speed range			No barred speed range			No barred speed range			9
10	No sepcial requirements (However 5 blade impeller must be avoided)			No barred speed range	No barred speed range		No barred speed range	—		No barred speed range	—		10
12	Countermeasure may be found necessary. Please consult engine builder			—	—		Barred speed range No barred speed range			No barred speed range No barred speed range			12

TABLE 12.3 CONTROL VALUES OF SOUND LEVELS PROPOSED FOR SWEDISH SHIPS

Space	Control Value dB(A)	dB(C)	Space	Control Value dB(A)	dB(C)
Engine Spaces			15. Crane and winch control space	85	
For existing ships			16. Deck office for port use	60	
1. Periodically Manned engine rooms	100		17. Bedrooms, living rooms, hospital background level with A/C off	$45 + k$	
2. Workshops and manned stores	80		Ditto with A/C on	$50 + k$	
3. Control rooms	70		18. Corridors	60	
For new ships			19. Offices	60	
4. Engine space with running engines	100		20. Dining rooms	60	
5. Engine space with engines stopped for overhaul	80		21. Galley, scullery, service room, etc	$60 + k$	
6. Workshops and manned stores	$70 + k$		22. Provision stores and similar	85	
7. Control rooms, noise recovery space, etc	65		23. laundry, washing room	70	
			24. Background level from equipment	70	
General			25. Sauna, dressing room	65	
8. Unmanned engine space	120		26. Outdoor recreation spaces	65	
9. Ditto with stopped engines	85		27. Remaining deck spaces	70	
10. Fan rooms, converter rooms, etc	85		28. Rooms for cultural activities	60	
			29. Conference room, library	50	
Other Spaces			30. Hobby room	70	
11. Wireless room	60	80	31. Gymnasium	60	
12. Navigating room, sea cabin	60	80	32. Passenger spaces max.	70	
13. Bridge wing, outdoor maneuvering place	60	80	Accepted Deviations		
			GRT > 60,000	$k = 0$	
			10,000 < GRT < 60,000	$k = 5$	
14. Fans during cargo handling	70		5,000 < GRT < 10,000	$k = 10$	
			1,000 < GRT < 5,000	$k = 15$	

Source: Flising (1978).

262 THE ENGINE AND ITS ENVIRONMENT

their status in 1978) for Swedish ships [Flising (1978)].* (Note that a correction k is included to allow higher levels for some spaces in smaller ships.) They are offered here as a typical example of the various noise stipulations that have been published.

Some typical source levels are given by Table 12.4. The table is also from Flising, but apparently his sources are data supplied by a diversity of manufacturers, so that all items may not be measured to the same standards. The velocity level figures are vibrational velocity magnitudes measured on the surfaces of the machines (not actually "sound," therefore); this is appropriate indication of the item's strength as a source of structure-borne sound.

A more general and qualitative view of machinery sound sources can be had from Figure 12.18, indicating that the main engine, propeller, generator engines, and exhaust stack outlet are the principal sources of sound from a low speed propulsion plant. (The stack outlet is obviously not a source in itself, but the principal location from which engine noise escapes to the surroundings.) Structure-borne noise, which is vibration that travels through structure to become noise at some remote point, is not indicated in the figure, but is important nonetheless.

The exhaust stack is perhaps the strongest source of engine noise, and must be provided with an attenuator ("muffler") within the exhaust line. Often the waste-heat heat exchanger is sufficient for this duty.

Within the machinery space, the turbocharger is the strongest source of airborne noise, with the noise being radiated from the turbocharger surfaces, from the surfaces of adjoining intake and exhaust ducting, and from the air intake opening. To combat this, engine builders insulate the surfaces, and fit sound baffles around intake openings. The curves of Figure 12.20 show measurements made by Sulzer on its RN90 (2900 bhp/cylinder) and RN90M (3500 bhp/cylinder) engines after thorough treatment of this nature.†

Now, as made obvious by Table 12.3, the major acoustic problem is not the noise level at the engine, but the results throughout the many spaces of the ship. The design problem is quite complex, for other sources (ventilation, for example) contribute, and sound may follow diverse paths

---

*The symbol dB in the table is the common "decibel" sound power level unit, it being 10 times the logarithm of sound power above the reference level of 10 to 12 W. The (A) and (C) modifiers describe frequency response characteristics of the measuring instrument. C, for example, has a virtually flat response over the sound frequency range 50 to 5000 Hz, while A has its major response in the 1000 to 10,000 Hz range.

†The dashed lines in Figure 12.20 are contours of equal loudness based on average human subjective response. For example, you are likely to think that a 1000 Hz sound of 50 dB and a 63 Hz sound of 75 dB are both of equal loudness (the 50 "phon" contour).

TABLE 12.4 TYPICAL SOUND SOURCE STRENGTHS IN A SHIP MACHINERY SPACE

Sound Source	Sound Power Level [dB(A)] (rel 10^{-12} W)	Sound Level Mean at 1 m Distance [dB(A)] (rel 2×10^{-5} Pa)	Velocity Level [dBCA] (rel 5×10^{-8} m/sec) At Engine Base	Velocity Level [dBCA] (rel 5×10^{-8} m/sec) At Engine Foundation Base
Slow speed diesel engines	130	104	80	70
Medium speed diesel engines				
> 2000 kW	125	104	90–80	95
< 2000 kW	122	104	—	—[a]
Diesel generators	115–122	97–105	85–90	65–70,[a] 80
Reduction gear for diesel machinery	95–100	75–83	85–90	95
Steam turbines	—	87–102	—	—
Turbo generators	106	85	—	—
Electric motors 15–45 kW 45–110 kW		<89 <93		
Gear wheel hydraulic pumps >300 l/min		80–90		
Screw wheel hydraulic pumps >300 l/min		75–80		75
Diesel-driven hydraulic pump units		—	115	
Water pumps	86–108	75–90	85	85[a]
Separators			75	65
Starting air compressor, 25 bar		85–90	75–87	65
Working air compressor, 7 bar	103–113	88–96		
Special performance compressor		<85		

Source: Flising (1978).
[a] = Vibration insulated.

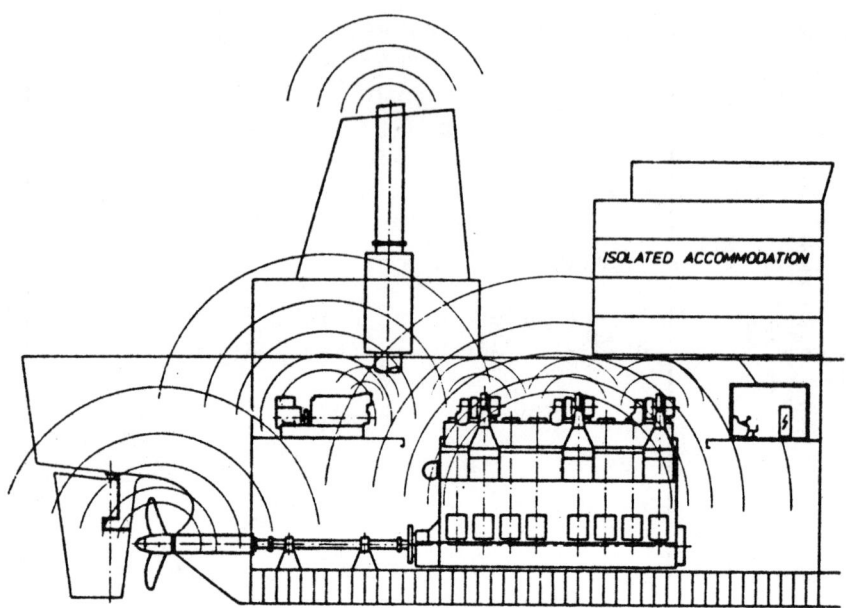

Figure 12.18 Principal sources of sound and vibration [from Smit (1979)].

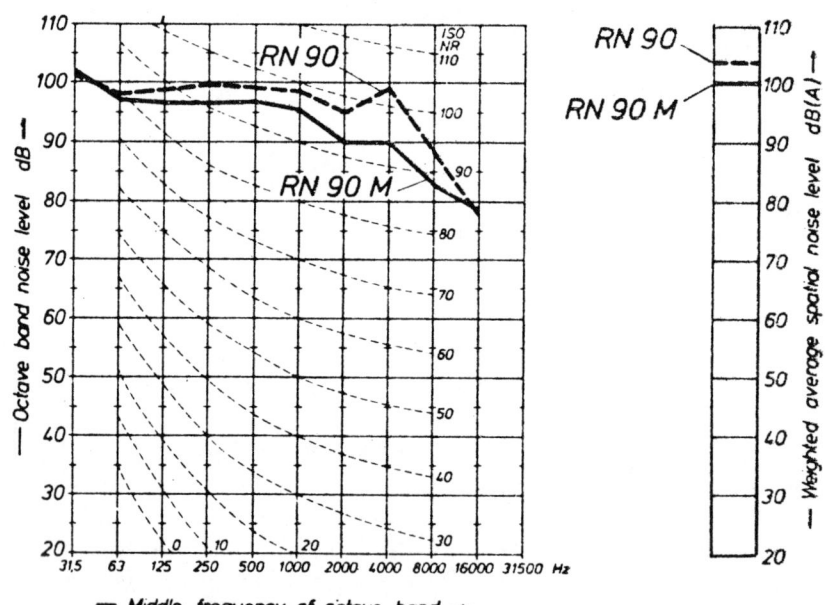

Figure 12.19 Airborne noise levels in vicinity of Sulzer RN90 and RN90M engines (Sulzer drawing).

THE ENGINE AND ITS ENVIRONMENT BEYOND THE SHIP   265

from source to point of interest. Theory and calculation are of scant help. We can only observe here that the trend is to insulate as far as feasible the major points of engine sound emanation, and as far as practicable to isolate sensitive areas from the engine. The latter measure is evident in Figure 12.18. This figure shows the accommodation structure ("deckhouse") completely separate from the exhaust stack—quite different from the traditional arrangement of an engine casing through the center of the accommodation area. It also indicates that the accommodation structure is separated from the main deck by resilient mountings. In such arrangements, piping and cabling coming from the machinery spaces are provided as well with flexible sections to thwart structure-borne sound and vibration.

12.10 THE ENGINE AND ITS ENVIRONMENT BEYOND THE SHIP

As a source of oil pollution, the diesel engine is essentially no source at all. Although the shipboard purification processes for fuel oil and lubricating oils produce an oily residue, its normal quantity is so small that it can be retained in a ship's slop tanks, or disposed of in an oily waste incinerator.

The engine cooling water does, naturally enough, reenter the sea at a higher temperature than that of the sea, but given the small magnitude of the discharge (compared to a shoreside power generating station, say), and its transient characteristic, the "thermal pollution" impact is trivial.

Atmospheric discharge of engine exhaust is the most likely source of nonnegligble environmental impact of a marine propulsion engine, for the products of incomplete combustion, the oxides of nitrogen, and the sulfur dioxide products that are the bane of several types of shoreside power sources are present to some degree at sea as well. However, as of 1980 no regulatory body had imposed any limitations on these exhaust constituents from marine engines, and this doubtless on the grounds that marine power plants do not occur in the massive concentrations of the swarming automobile, nor in the massive size of electric power generating stations. The only legal concern for the engine builder and marine engineer is for the visible pollution—smoke, that is—regulations of some port communities. The engines are universally designed to operate below an acceptable "smoke limit," and if maintained in good order, do not violate these pollution stipulations.

But what if exhaust emission regulations comparable to those applied to automobiles were to invade the sea? The literature of automotive engineering* treats diesel exhaust properties in great detail, since the diesel is also an automotive and truck engine in widespread use. In general, the diesel faces difficulty only in meeting desired low levels of the nitrogen oxides. The same is true for the low speed marine engine, but its problems are milder because its conditions (principally its low speed) for combustion are the best of all diesels. The oxide of nitrogen level can be reduced by adjustments of running conditions (for example, changing fuel injection timing), and according to Smit [Smit (1979)] the only penalty on engine performance would be in limits on mean effective pressure, which would halt progress toward higher specific engine outputs, but otherwise not compromise the virtues of this engine type.

Burning of high sulfur fuels may suggest limits on sulfur dioxide. To my present knowledge, the question of regulation has not been raised, nor has the question of what countermeasures would be needed or feasible.

12.11 REFERENCES

1. Den Hartog, J P (1956), *Mechanical Vibrations,* fourth edition, McGraw-Hill.
2. Flising, A (1978), "Noise Reduction in Ships," *Transactions,* Institute of Marine Engineers, vol 90, series A, pages 292-320.
3. Konsted, E (1975), "Engine Plants in Triple Screw Container Ships," *Transactions,* Institute of Marine Engineers, Vol 81, series A, pages 144-156.
4. Long, C L (1971), "Propellers, Shafting, and Shafting System Vibration Analysis," *Marine Engineering,* Society of Naval Architects and Marine Engineers.
5. Salzman, R H and Pamidi, P R (1973), "Machinery Vibrations in Marine Systems," Marine Engineering Society in Japan, marine engineering symposium, Tokyo,
6. Smit, J A (1979), "The Marine Diesel: Recent Design Developments," *Marine Technology,* vol 16,2, pages 119-235.
7. Spaetgens, T W (1961), "Torsional Vibrations—Some Actual Marine Problems and Solutions," *Transactions, Institute of Marine Engineers Canadian Division Supplement.*

*See, in particular, the publications of the Society of Automotive Engineers.

8. Wilson, W K (1956), *Practical Solution of Torsional Vibration Problems*, second edition, John Wiley & Sons, New York.

12.12 NOTATION FOR CHAPTER 12

dB	decibel
E_E	engine damping energy
E_I	hysteresis damping energy
E_p	propeller damping energy
J	propeller advance coefficient
J_e	polar mass moment of inertia of engine component
K_e	numerical coefficient (engine damping)
K_Q	propeller torque coefficient
l	stroke length
MAN	Maschinenfabrik Augsburg Nurnburg
Mn	magnitude of torque harmonic at cylinder n
m_{rec}	reciprocating mass
m_{rot}	rotating mass
mm	millimeters
Q	propeller torque
r	crank radius
r	resultant (in Figure 12.16)
rdn	radians
rpm	revolutions per minute
t	time
α	numerical coefficient
β	torsional vibration amplitude ratio at cylinder n
ϕ_n	phase angle at cylinder n
Θ_n	amplitude of vibration at any cylinder
Θ_p	amplitude of propeller vibration
Ω	speed of propeller rotation
ω	vibrational frequency

INDEX

Acid in combustion products, 134-135
Additives to fuel, 137
Advance coefficient of propeller, 67
Air/fuel ratio, 84
Air pumps, 100-101
Air supply, 49
Ambient conditions, 60
American Bureau of Shipping rules, 52, 163, 195
Arrangement of machinery, 235-236
Ash in fuel, 128
Auxiliary power drive, 208
Auxiliary systems, 179-213
Axial vibration, 255

Bearings, 34
Blow-by, 165
Bore/stroke ratio, 20
Brake power, 111

Carbon contamination, 135
Carbon residue, 127
Cetane number of fuel, 129
Characteristics:
  engine, general, 57-58
  fuel consumption, 55
  torque and power
    engine, 15-17, 55
    propeller, 69-71
Classification of engines, 3-4
Classification society rules, 52, 163, 194-196
Cleaning, 210, 224

Cloud point of fuel, 127
Coefficients:
  advance, 67
  thrust, 67
  torque, 67
Comparison criteria, 118
Compression ratio, 13
Constant pressure turbocharging, 97-100
Consumption of lubricant, 59, 194
Contamination of fuel, 127-129
Control:
  computer, 147-148
  controllable pitch propeller, 167-169
  directional, 151-152
  electronic fuel control, 147-149
  fuel setting, 155
  gaseous fuel, 169-170
  governing, 157-162
  maneuvering, 155-156
  remote, 149-151
  safety, 156
  speed control vs. fuel control, 146
  starting, 146-147, 155
Controllable pitch propeller, 75, 167-169
Cooling:
  fuel injectors, 181
  general, 51-52
  jacket, 181-183
  piston, 183
Crankcase explosion, 218
Crankshaft, 35-36

INDEX

Cylinder:
  block, 37
  covers, 39
  liners, 37-38
  lubrication, 137-138
  ports, valves, 40-45

Design tasks, 6-8
Diesel cycle, 12-13
Diesel oil, 130
Differential thermal efficiency, 119-120
Dirt in fuel, 128
Displacement, 19
Distillation of seawater, 183-185
Distillation temperatures of fuel, 126

Effective power, 66
Efficiency:
  hull, 66
  mechanical, 15
  scavenging, 85
Electric load, 204-206
Electronic fuel control, 49-50
Engine characteristics, 57-58
Engine mounting, 236-246
Environmental problems, 265-266
Environment for engine, 234-267
Exhaust, 49, 198-200
Exhaust gas temperature, 61
Expected maintenance, 219-233
Explosion in crankcase, 218

Failures, 215-218
Fire, 218
Fire protection, 208
Flash point, 126
Fresh water distillation, 183-185
Friction torque and power, 15-16
Fuel:
  additives, 137
  consumption characteristics, 55
  contamination, 127-129
  filtering, 136
  flash point, 126
  gaseous, 138-140
  heating value, 127
  homogenizing, 137
  ignition quality, 129

  injector cooling, 181
  injectors, 45
  pour point, 127
  properties, 125-143
  pumps, 45
  system, 188-191
  treatment, 136-137
  viscosity, 125, 132
  washing, 136
Fuels, 123-143

Gaseous fuel, 138-140, 191
Gaseous fuel control, 169-170
Gas oil, 130
Governors and governing, 157-162, 171-178

Heat balance, 60
Heating value of fuel, 127
Homogenizing of fuel, 137
Hull efficiency, 66
Hull resistance, 66
Hull service margin, 114

Ignition quality, 129
Indicator diagram, 15
Inertia force, torque, 256-258
Injection, 10
Injectors, 45
In situ repairs, 231

Jacket water cooling, 181-183

Lateral vibration excitation, 258-259
Liner, cylinder, 37-38
Loop scavenging, 87-88
Lubricant consumption rate, 59, 194
Lubricants, 137-138
Lubricating systems, 191-194
Lubrication, 51

Machinery arrangement, 235-236
Machinery weight, 206-207
Maintenance, 219-233
Maintenance equipment, 225-230
Maneuvering control, 155-156
Marine diesel oil, 130
Marine gas oil, 130
Matching engine and propeller, 71-75

Mean effective pressure, 13-14, 20, 22
Mechanical efficiency, 15
Monitoring, 162-167

Noise, 259-265
Non-propulsive loads, 78-80

Operating margin, 114

Piston:
  cooling, 32-33, 183
  rings, 32-33
  speed, 18
Pistons, 27-33
PLAN formula, 14-15, 18
Pour point, 127
Power:
  brake, 111
  characteristics:
    of engine, 15-17, 55
    of propellers, 69-71
  effective, 63, 66
  friction, 15
  shaft, 111
  specific, 119
Propeller characteristics, 69-71
Propeller/engine transients, 80
Propeller matching to engine, 71-75
Properties of fuels, 125-143
Properties of lubricants, 137-138
Pumps:
  fuel, 45
  lubricant, 191-194

Rating corrections, 112
Rating of engine, 108-122
Remote control of engine, 149-151
Repair *in situ*, 231
Residual fuel oil, 130-132
Resistance of hull, 66
Rings, piston, 32-33
Running gear, 27-33

Scavenging, 11, 84-107
  cross, 87-88
  efficiency, 85
  loop, 87-88
  ratio, 86
  uniflow, 87-88

Scuffing, 165
Service margin, 114-116
Spare parts, 230
Specific power, 119
Speed of piston, 18
Speed rating, 109-110, 114
Starting, 146-147, 155, 194-196
Stroke length, 23-24
Structure of engine, 39-40
Sulfur in fuel, 127, 135
Surge of turbocharger, 94-95

Thrust:
  coefficient, 67
  deduction, 66
  power, 65-66
Torque:
  characteristics:
    of engine, 15-17, 55
    of propeller, 69-71
  coefficient, 67
Torsional vibration, 246-255
Total base number (TBN), 138
Transients, 80
Treatment of fuel, 136-137
Turbocharger:
  characteristics, 94
  surge, 94-95
Turbocharging, 84-107
  constant pressure, 97-100
  pulse, 97-100
  two-stage, 104-106

Uniflow turbocharging, 87-88

Vanadium in fuel, 127-128
Ventilation, 196-198
Vibration:
  axial, 255
  torsional, 246-255
Viscosity, 142-143
  of fuel, 125, 132

Wake fraction, 66
Washing of fuel, 136
Waste heat use, 200-204
Wear, 216, 219-224
Wear detection, 163-165
Weight of machinery, 206-207